Galileo

Past Masters

AQUINAS Anthony Kenny
BACON Anthony Quinton
BURKE C. B. Macpherson
DANTE George Holmes
GALILEO Stillman Drake
HOMER Jasper Griffin
HUME A. J. Ayer
JESUS Humphrey Carpenter
MARX Peter Singer
PASCAL Alban Krailsheimer

Forthcoming

ARISTOTLE Jonathan Barnes
AUGUSTINE Henry Chadwick
BACH Denis Arnold
BAYLE Elisabeth Labrousse
BERGSON Leszek Kolakowski
BERKELEY J. O. Urmson
JOSEPH BUTLER R. G. Frey
CARLYLE A. L. Le Quesne
COBBETT Raymond Williams
CONFUCIUS Raymond Dawson
COPERNICUS Owen Gingerich
DARWIN Jonathan Howard
DIDEROT Peter France
ENGELS Terrell Carver
ERASMUS James McConica
GIBBON J. W. Burrow
GODWIN Alan Ryan
GOETHE J. P. Stern
HEGEL Peter Singer
HERZEN Aileen Kelly
JEFFERSON J. P. Greene
LAMARCK L. J. Jordanova
LINNAEUS W. T. Stearn
LOCKE John Dunn
MACHIAVELLI Quentin Skinner
MALTHUS Gertrude Himmelfarb
MILL William Thomas
MONTAIGNE Peter Burke
THOMAS MORE Anthony Kenny
MORRIS Peter Stansky
NEWMAN Owen Chadwick
NEWTON P. M. Rattansi
PLATO R. M. Hare
ROUSSEAU John McManners
ST PAUL Tom Mills
SHAKESPEARE Germaine Greer
ADAM SMITH A. W. Coats
TOLSTOY Henry Gifford
and others

Stillman Drake

GALILEO

Oxford Melbourne Toronto
OXFORD UNIVERSITY PRESS
1980

Oxford University Press, Walton Street, Oxford OX2 6DP
*London Glasgow New York Toronto
Delhi Bombay Calcutta Madras Karachi
Kuala Lumpur Singapore Hong Kong Tokyo
Nairobi Dar es Salaam Cape Town
Melbourne Wellington
and associate companies in
Beirut Berlin Ibadan Mexico City*

First published as an Oxford University Press paperback 1980 and simultaneously in a hardback edition

British Library Cataloguing in Publication Data
Drake, Stillman
Galileo. - (Past masters).
1. Galilei, Galileo
2. Astronomers - Italy - Biography
509'.2'4 QB36.G2
ISBN 0-19-287527-2
ISBN 0-19-287526-4 pbk

Printed in Great Britain by
Cox & Wyman Ltd, Reading

Preface

The silencing and punishment of Galileo toward the end of a life devoted to scientific inquiry was an event of profound significance for our cultural history. Its full understanding requires much more than an assumption of inevitable conflict between science and religion – a cliché which originated largely from the case of Galileo that it is widely used to explain. If any simple explanation existed, it would rather be in terms of the customary ruthlessness of societal authority in suppressing minority opinion, and in Galileo's case with Aristotelianism rather than Christianity in authority. To understand the fate of Galileo requires knowledge of events throughout his whole career coupled with that sense of inevitable outcome, hidden from the actors, that authors of classical Greek tragedy gave to its spectators.

Those spectators were kept posted by the chorus, whose role I must undertake to play, since in a short book it is not possible to deal in detail with more than a single theme in Galileo's multiple activities. I have chosen as the focal point Galileo's condemnation by the Roman Inquisition in 1633, with his biography as the background. That choice entails certain limitations. There is not room to justify statements about technicalities of Galileo's science, even when they depart from currently received scholarly opinion; all I can do is to assure the reader that they are based on documentation more extensive than general histories of science have used.

Galileo's later influence on science, and his intellectual origins – his relationship to philosophy and his debt to medieval science – are questions that are of course important to philosophers of science and historians of ideas, but to dwell on them in a biography would so interrupt the narrative as to destroy its continuity. Accordingly I shall in this book touch on such matters only to the extent I consider useful to readers or necessary to explain my personal view of Galileo's science and its relation to philosophy.

Galileo's science was not that of Descartes or Newton, the

two thinkers who were most influential in shaping its development after his death. Neither was it that of the universities of his own day, which had developed mainly from the natural philosophy of Aristotle. Galileo's physics was founded on his own actual measurements, which, through ingenuity and precision, led him to his law of falling bodies. That was not a medieval approach to the study of motion. Neither was it a philosophical one, because natural philosophy sought causes not laws, and causes are revealed not by measuring but by reasoning. 'What has philosophy', asked Galileo in 1605, 'got to do with measuring anything?'

His approach was scientific rather than philosophical because measurement remains always approximate, however much its instruments and procedures are refined. Philosophers seek exact knowledge, not just better and better approximations such as satisfy scientists. When Galileo began making physical measurements, he put philosophy aside temporarily; and when measurement led him to laws he lost interest in causes, indefinitely postponing his return to philosophy.

A law found by measurement is necessarily mathematical in form, whence its manipulation by proportionalities will reveal consequences no less certain to be borne out by measurement. In that pursuit Galileo came to see mathematics as essential to physics – not because the paper world of mathematics is more interesting than is the sensible world around us, but because the language of mathematics enabled him to read that 'great book of nature', as he called it.

My purpose is to present briefly the progress of Galileo's thought as it matured and the growing oppositions to it that he strove to counter as effectively as he could. In so doing, I shall cite many passages from his books in English translation. To avoid superfluous footnotes, each quotation is followed by a key and page numbers, referring to a book briefly identified opposite. Full titles will be found in the Reading list at the end of this volume. Grateful acknowledgement for use of copyrighted material is made to the publishers.

D	Galileo, *Dialogue . . .* (University of California Press, Berkeley)
D&O	S. Drake, *Discoveries and Opinions of Galileo* (Doubleday & Co., New York)
GW	S. Drake, *Galileo At Work* (University of Chicago Press, Chicago)
L	B. Dibner and S. Drake, *A Letter from Galileo* (Burndy Library, Norwalk, Conn.)
OP	A. Favaro (ed.), *Opere di Galileo* (G. Barbera, Florence)
PLG	[M. Allan-Olney], *The Private Life of Galileo* (Macmillan, London)
TNS	Galileo, *Two New Sciences* (University of Wisconsin, Madison)

Contents

To the memory of

Bryant S. Drake *Chemical Engineer*

Raymond C. Brooks *Pastor*

Introduction

The earliest explanations of man and his universe appear to have arisen from religion. Philosophy, in western society at any rate, came later on the scene, and last of all, science. It is accordingly natural that philosophy should at first have been guided by religion and have guided science. Certainly that was the pattern in European culture from the revival of learning in the twelfth century to the time of Galileo.

Galileo's significance for the formation of modern science lies partly in his discoveries and opinions in physics and astronomy, but much more in his refusal to allow science to be guided any longer by philosophy. By stages, his rejection of the long-established authority of philosophers induced them to appeal to the Bible for support, and there ensued a battle for freedom of scientific enquiry which profoundly affected the development of modern society.

Galileo's role in that battle is widely supposed to have been that of hurling a defiant challenge to religious faith in the name of science. That was by no means his intention, though it is true that theologians proceeded to nip Galileo's science in the bud, which may not have been their intention at the outset. Galileo's science entered only indirectly into the celebrated event on which it is my hope in this book to shed new light; that is, the trial and condemnation of Galileo by the Roman Inquisition in 1633.

Eminent scholars in all the principal European nations, examining that event for over a century, have still not reached unanimity about it. Even scholars have tended to divide between the two camps of science and faith, perhaps because in an adversary trial procedure there are only two possible verdicts – 'guilty' or 'not guilty'. In such situations it is tempting to conclude for one side or the other despite lingering puzzles. Much has been done to reduce the number and the seriousness of those puzzles, with growing mutual respect between leading representatives of both the religious and the scientific communities. Only by considering every possibility is there real hope of establishing historical truth in a complex situation. The balance of probabilities is greatly altered by every new hypothesis. In taking a position different from any I have read, I do not mean to disparage other solutions of this complicated problem, but only to offer a hypothesis which, however outlandish it may appear at first sight, may remove old puzzles without creating new ones just as troublesome.

That is what is often done in science itself, and by an amusing coincidence this is illustrated by the same scientific hypothesis that gave rise to the dispute in which Galileo became involved. For when Copernicus proposed holding the sun fixed and moving the earth, astronomers had long been able to make calculations of past and future planetary positions as accurate as those of Copernicus, and they had come to accept as irreducible such astronomical puzzles as the apparent dependence on the sun of the periods of the inner planets, and the stopping and turning of the outer planets when in opposition to the sun. The Copernican hypothesis did not solve all the problems of astronomy, but it did remove some ancient puzzles that had distracted attention from what may be called the serious work of astronomers. All the same, the Copernican hypothesis seemed outlandish, because anyone could *see* that the earth does not move.

Just so, my hypothesis about the Galileo affair may seem at first highly improbable. It is that Galileo was a zealot not for the Copernican astronomy, but for the future of the Catholic Church and for the protection of religious faith against *any* scientific discovery that might be made. To those who suppose this unthinkable, I can say at present only that anyone willing to accept it merely as a fiction will be able, by further reading, to see how many long-accepted problems can be made to vanish before his very eyes – much as Galileo tells us that a certain professor at Pisa, Antonio Santucci, set out to study Copernicus in order to refute him, and was instead won over to his views. That same thing had probably happened earlier to Galileo himself, since we all know that no one is born believing that the earth moves, and few do not argue against that idea at first.

Nevertheless there is a widespread belief that Galileo, without adequate scientific evidence, battled from his earliest years for the Copernican system. If that were true, it would be hard indeed to understand his very cautious approach to other scientific problems. Even harder to explain would be the fact that he never mentioned his preference for the Copernican astronomy until he was over thirty years of age, and then remained silent about it for another decade. By the time Galileo endorsed the new system in print, he was nearly fifty, and he had meanwhile done a great deal of solid work in both physics and astronomy. Those who think of Galileo as a Copernican zealot are unfamiliar with what Leonardo Olschki called 'his scientific personality'.

The picture of Galileo's personality presented year by year in his correspondence and publications is that of a prudent man, not given to forming conclusions without having weighed the evidence, well aware of social customs, and disinclined to quarrel with highly placed persons in Church or state. Most of the controversies in which he engaged were precipitated by others attacking his constructive views, not the other way round; and to most such attacks he did not even reply. Now, it is conceivable that a man of fifty might suddenly become an unreasoning zealot for some cause, without his having previously given evidence of instability; but that would be most unusual in the case of one who had formed and preserved friendships with many persons of good intelligence, sound judgement, and quite varied positions and backgrounds.

The common assumption that Galileo was a Copernican zealot has resulted in sharply conflicting pictures of his character and personality. I do not say that that shows it to be false, though in logic only false assumptions lead to contradictory conclusions. In one picture, Galileo was an intuitive hero of science who, without sufficient evidence, did battle against benighted tradition; in another, he was an irresponsible troublemaker who injured the cause of true science by taunting the forces responsible for maintaining social order. At least one of those pictures must be wrong and should be discarded; I believe that both should be. Instead of deducing Galileo's personality from his confrontation with the Inquisition, we ought to study it independently for the light it could throw on that conflict. There are abundant documents from which to gauge Galileo's attitudes and characteristics by his words and actions in a wide variety of affairs. Most people are fairly adept at judging the character and personality of others, or at least at distinguishing unreasoning zealots from men of good sense.

Perhaps it is considered unscholarly to suggest personality as a consideration in weighing alternative hypotheses about the trial of Galileo, as a concept too subjective for scholarly research. Some, at least, prefer to regard Galileo more as a puppet of great social and intellectual forces than as a human being capable of thinking for himself, or of deciding anything except on the basis of some elusive philosophy. I do not say that personality is an objective concept, but to me it is no less a proper topic of research than is the Catholic Church, to which many scholars have devoted many studies. To me, the Catholic Church was composed of a great many individuals, each as complex as Galileo, among them three cardinals (out of ten responsible) who declined to sign the sentence passed against him, and an archbishop who invited Galileo, during his troubles, to be his guest when the trial was over, and who, immediately after it, saved his sanity and probably his life. Each individual in the Catholic Church with whose writings and actions I am familiar has for me a personality no less understandable and consistent than Galileo's – and no more so. But the Catholic Church as a single entity, then as now, lies entirely beyond my comprehension. Its sheer durability strikes me as the most baffling social phenomenon that exists, and far less easy to study than the personality of Galileo.

The personality of an individual is carefully considered by those who might be endangered by his conduct. Two major Italian universities and two grand dukes of Tuscany put their trust in Galileo over many years, and the trust of universities and heads of state has never been easy to win, especially in Italy. Clearly he was not regarded as a troublemaker by responsible employers, though they knew well enough that he was a pugnacious fighter. Their judgement that he fought only for just causes, expressed by their employment of him, contributes to our knowledge of his personality.

Long study of Galileo's writings has convinced me that he chose words with care and that they reflect his sincere convictions. In one matter, however, I reserved judgement. Italian usage, then as now, requires courteous phrases and some kinds of exaggeration that once flourished also in English and that might be read as not sincere. It is also true that Catholic practice required expressions of deference to Church doctrines and dignitaries which may not always have been felt. In learning to read Galileo's Italian, I tried to avoid mistaking polite conventions for heartfelt utterances. For that reason I long remained unimpressed by, and indeed indifferent toward, Galileo's frequent avowals of zeal for the Church. He seems never to have used the word 'zeal' except in that connection. It is a strong word, unnecessary to use at all, whence it is likely to appear in only two circumstances – heartfelt zeal, and the defensive manoeuvring of one who, far from feeling it, protests too much.

It was only while writing this book, and in fact after having written some of it rather differently, that it occurred to me, quite suddenly, to try the hypothesis that Galileo had spoken not conventionally but sincerely of his zeal for the Church, and that Catholic zeal may indeed have motivated him in taking certain risks for which he was ultimately not rewarded but punished. Having previously read many times the relevant documents, I had them, so to speak, simultaneously present, together with Galileo's words on various related occasions. The effect on me of this new hypothesis was electric, like happening on a neglected document that resolves old perplexities. If Galileo's chief concern was for his Church, and he saw it on the brink of taking a fatal misstep urged by old foes of his science, then the grand duke's trust in him against the advice of a seasoned Roman ambassador ceases to be puzzling. Galileo's employer would approve of Catholic zeal where he would have judged Copernican zeal just as did his ambassador – to be merely misguided and rash. That three cardinals of the Inquisition should refuse to sign the sentence against Galileo seems odd – unless they were personally certain of his devout Catholicism. Many former perplexities of the whole affair vanished like these under the new hypothesis, which I suppose never occurred to Catholic partisans in old scholarly debates because Galileo seemed to have defied a Church edict, nor to partisans of cold science to whom religious zeal seemed out of character for Galileo.

There is little hope that any new document concerning this matter will be discovered; in a way, the difficulty has long been to reconcile all the documents we have. Looking at them only in the light of later events (for they were not published till a century ago), historians took many of Galileo's statements as insincere because the wide breach between religion and science had come to be accepted as a matter of fact. The Darwinian controversy which was raging when Galileo's trial documents were finally published probably affected their historical interpretation.

Yet before the Galileo affair there had been neither a breach between religion and science nor any distinction between science and philosophy. Galileo created the science that could not be accepted by philosophers, which is why all recent attempts to make a philosopher out of him have created much heat but no light. But it was not Galileo who created the breach between religion and science. As Galileo explicitly said in his *Letter to Christina* at the outset of the Copernican battle in 1615, that breach was invented by professors of philosophy:

> They have endeavoured to spread the opinion that such [Copernican] propositions in general are contrary to the Bible and are consequently damnable and heretical. They know that it is human nature to take up causes whereby a man may oppress his neighbour, no matter how unjustly. ... Hence they have had no trouble in finding men who would preach the damnability and heresy of the new doctrine from the very pulpit ...
>
> Contrary to the sense of the Bible and the intention of the Church Fathers, if I am not mistaken, they would extend such authorities until even in purely physical matters, where faith is not involved, they would have us altogether abandon reason and the evidence of our senses in favour of some biblical passage, though beneath the surface meaning of its words this passage may contain a different sense. (D&O 179)

Galileo did not even blame the priest – a young Dominican firebrand who hoped (mistakenly, as it turned out) thus to advance his career in the Church – for denouncing him from the pulpit in Florence. For that he blamed human nature, not religion. Behind the action of the foolish priest were the professors of philosophy who undertook to interpret the Bible and create a new heresy. Galileo would not absolve them from blame for resorting to power when reason went against them. By their own principles, reason should prevail in everything.

The charge laid by Galileo against philosophers was the utilisation of a frailty in human nature, betraying their own principles. They alone were responsible for introducing the Bible into their dispute with Galileo, which he sincerely regarded as an impious action on their part.

It is curious that in the enormous literature which has grown out of the events, Galileo's charge against the professors of philosophy has not even been noticed. One might think them to have been innocent bystanders at a confrontation which did not concern them, or at worst clownish reactionaries who wrote some trifling books in opposition to the new science of Galileo. The documents, however, show that Galileo's charge was just; before any priest spoke out against him, his philosopher opponents declared his opinion contrary to the Bible and considered enlisting some priest to say so publicly, for which they were rebuked by a churchman. Galileo knew of those events, and who his enemies were behind the scenes, before he wrote a word about the relation of science to religion.

1 The background

Dante called Aristotle 'the Master of those who know'. Aristotle was so regarded by learned men from the time of Aquinas to that of Galileo. If one wished to *know*, the way to go about it was to read the texts of Aristotle with care, to study commentaries on Aristotle in order to grasp his meaning in difficult passages, and to explore questions that had been raised and debated arising from Aristotle's books. University education had been patterned on those procedures from its very beginning in the thirteenth century. As Aristotle had lived before the Christian era, he was recognised to have been mistaken on some points, but they were not many and theologians had found and corrected them. Aristotle was commonly referred to as the Philosopher, with a capital P; all matters of knowledge belonged to Philosophy just as all matters of faith belonged to Sacred Theology.

Physical science in general constituted 'natural philosophy'; that is, knowledge of Nature, which was *physis* in Greek. Aristotle had covered this in several works, notably in his *Physics, On the Heavens, Meteorology*, and books about the creation and the coming to an end of things. The principles of physical science were determined in Aristotle's *Metaphysics*, written after he had composed his books on science, since it would not be proper to have science discuss its own principles, and still less to build it on arbitrary principles determined in advance without careful study of nature.

The general pattern of Aristotelian physics and cosmology is probably known to most readers, and a brief summary cannot do it justice. Nevertheless, in order to make clearly perceptible the sources of the opposition against which Galileo had to contend when he suggested a different approach to the study of nature, a short sketch of the established approach as taught in the universities of his time will be attempted.

Basic to Aristotle's philosophy was the goal of understanding why things are as we find them, why they could not be (or have been) otherwise, and why it is best that they are as they are. To understand those necessities it is essential to penetrate to the causes of things, and to grasp the ultimate purpose behind all events in nature. Building on the work of his predecessors, Aristotle adopted as fundamental the four 'elements' – earth, air, water, and fire – and four qualities in paired opposites – heat and cold, moisture and dryness – as associated with them. Natural places to which the elements belonged were assigned, and natural tendencies of gravity and levity by which they always strove to return to their natural places if removed therefrom. Logical rules were established by which causes could be determined for the effects we perceive in nature, causes being given only by reason and not presented to us directly by our senses. Substance, form, agent, and purpose were recognised as determining, or predominating in, causation of distinguishable kinds. The inner essences of things were identified in their definitions, and distinguished in that way from accidental properties they exhibited under various circumstances. Natural philosophy then consisted of causal explanation of observed phenomena in nature within such a logical and schematic programme.

The physics of Aristotle was concerned primarily with change, which he put as the most fundamental characteristic of nature, saying that to be ignorant of change was to be ignorant of nature. The term used by Aristotle for change was translated into Latin as 'motion', which eventually became restricted to what Aristotle called 'locomotion' (change of place with respect to time), recognised by him as somehow logically prior to other kinds of change, or always involved in or implied by it, but not of exclusive interest in his physics. Change of quality, as when iron changes from brown to red and then to orange and then to yellow as it is heated, or change by growth with the passage of time, was of equal interest to Aristotle.

Turning from physics to cosmology, Aristotle divorced the heavens from the elemental parts of the universe with the earth as their centre and the fire as the highest sphere, bounded by the sphere of the moon. Beyond the four elements, everything consisted of a fifth substance, the quintessence, which unlike them was not subject to any kind of change except locomotion, uniformly and in perfect circles. The cosmology of Aristotle appears to have been developed directly from a suggestion of

Plato's, mathematically rationalised by Eudoxus, so in this matter there was not a conflict between two philosophies. Aristotelian cosmology survived unaltered by Ptolemaic astronomy, despite its eccentric planetary orbits and the epicyclic motions which already strained the original simple notion of uniform circular motion around the earth. It could not survive the Copernican astronomy, which put the earth itself in motion against the basic principles of Aristotelian natural philosophy. Either the latter or the Copernican system had to be abandoned or altered beyond recognition.

Between 1605 and 1644 a series of books appeared in rapid succession in England, Italy, and France which laid waste the Aristotelian natural philosophy of the universities. The authors were Francis Bacon, Galileo, and René Descartes. The only conspicuous matter of agreement among them was that Aristotelian natural philosophy was not good science. Of course there had been earlier disparagements of the Philosopher and other programmes for improvement of science, especially during the sixteenth century, and accelerating toward its end. In the years named, others continued to appear. But for our present purposes it suffices to note a significant epoch in the cultural history of Europe marked by the appearance of three celebrated thinkers during a single generation, in three different countries, after four centuries of Aristotelian authority in science, all of whom opposed that on solid, though widely different, grounds.

Since we are here concerned only with the contributions of Galileo, it is appropriate to stress one significant way in which those differed from contributions made by Bacon and Descartes. Those two men are remembered for their philosophies, still called 'Baconian' and 'Cartesian'. Hardly any later philosopher in Europe or America, or any historian of philosophy, has ignored them. Galileo, in contrast, is remembered only for his contributions to science. Hardly any later philosopher or historian of philosophy took note of him, though several scientists and nearly all historians of science did.

The epoch in cultural history marked by Bacon, Galileo, and Descartes is usually called the Scientific Revolution, or by some its beginning. The seventeenth century was characterised by the rise of useful science as distinguished from science for its own sake, though that pursuit of course continued. Utility had been deliberately excluded from Aristotelian natural philosophy. Aristotle had nothing against practical knowledge, which he called *techne*; he simply did not consider it to be the same kind of thing as scientific knowledge, which he called *episteme*. From *techne* we have the word technology, which means to us largely the application of scientific knowledge, while from *episteme* we have the word epistemology, a branch of philosophy which deals with the theory of knowledge, scientific or any other. For Aristotle, however, the difference between *techne* and *episteme* was not a difference between application and theory, but was one of sources of knowledge and goals of knowledge. The source of technical knowledge was practical experience and its goal was, roughly speaking, knowing what to do next time. The source of scientific knowledge was reason, and its goal was the understanding of things through their causes.

The Scientific Revolution consisted to a large degree in erasing those classical distinctions and in bringing the kind of knowledge acquired from practical experience together with the kind achieved through reason, even at the cost of accepting knowledge of what to do next time in place of an understanding of the causes of things. This latter move is more politely described as the search for laws instead of causes.

Even now it appears to be offensive to philosophers to slight in any way the search for causes. That was still more offensive to them at the beginning, after centuries during which the whole purpose of science, or natural philosophy, had been to determine the causes of things. Descartes, for example, considered any contrary procedure so outlandish that he rejected Galileo's science out of hand because it had not started with an investigation of the causes of motion and of heaviness.

Aristotle's physics offered material, formal, efficient and final or purposive causes of every kind of change in nature. Galileo's physics dealt with local motion for the most part, and even then only with local motions of heavy bodies on or near the surface of the earth; and moreover did not attempt any causal explanation even of these. Not only did it fail to come to grips with most of the problems which concerned contemporary philosophers, but it contradicted express statements by Aristotle concerning (for example) speeds of fall of heavy bodies, and offered no causal explanation in place of his. It is clear why these philosophers regarded Galileo's science as beneath contempt: to them it appeared pitifully trivial and inadequate.

At the beginning of his career, of course, Galileo had attempted to investigate motion by causal reasoning, as he had been taught to do at the university. But when, late in life, he introduced his law of fall – the law that was to become the cornerstone of a new physics – he had this to say on the subject:

> The present does not seem to me to be an opportune time to enter into the investigation of the cause of the acceleration of natural motions, concerning which various philosophers have produced various opinions ... Such fantasies, and others like them, would have to be examined and resolved, with little gain. (TNS 158–9)

We shall see how, when, and perhaps why Galileo in due course abandoned causal reasoning, though he by no means gave up using the word 'cause', which is very useful when sensibly applied. But before we embark on his biography, it will pay us to glance at some of his mature remarks about science and its relation to philosophy.

Galileo's science was not a closed system, as was Aristotle's. It was not so much a collection of conclusions as it was a method. To the extent that it embodied conclusions, those were both piecemeal and incomplete, and they were expected by Galileo to remain so, no matter how far science progressed. In his own words in *The Assayer* of 1623:

> To put aside hints and speak plainly, and dealing with science as a method of demonstration and reasoning capable of human pursuit, I hold that the more this partakes of perfection, the smaller the number of propositions it will promise to teach, and fewer yet will it conclusively prove. Consequently the more perfect it becomes, the less attractive it will be and the fewer will be its followers. On the other hand [books with] magnificent titles and many grandiose promises attract the natural curiosity of men, and hold them forever involved in fallacies and chimeras without ever offering them one single sample of that sharpness of true proof by which the taste may be awakened to know how insipid is its ordinary fare. (D&O 239–40)

The ordinary fare to which Galileo referred was Aristotelian natural philosophy, which was a complete array of conclusions about physics and astronomy, marshalled under metaphysical principles and logical procedures which enable one to find the cause of any effect in Nature that might ever come up. Galileo's remark that as science progressed it would venture ever fewer propositions did not mean absolutely fewer, but fewer in comparison with natural philosophy and its grandiose programme of explaining everything that might be discovered. In contrast, he said:

> There is not a single effect in Nature, not even the least that exists, such that the most ingenious theorists can ever arrive at a complete understanding of it. This vain presumption of understanding everything can have no other basis than never understanding anything. For anyone who had experienced just once the perfect understanding of one single thing, and had truly tasted how knowledge is attained, would recognise that of the infinity of other truths he understands nothing. (D 101)

It might seem inconsistent for Galileo to say in one sentence that even the tiniest event in Nature will never yield to complete understanding, and then in another to imply that someone had understood some one thing perfectly and thereby realised the implications of that knowledge. The apparent inconsistency was resolved in this way:

> I say that the human intellect does understand some propositions perfectly, and thus in these it has as much absolute certainty as has Nature herself. Those are of the mathematical sciences alone; that is, geometry and arithmetic, in which the Divine intellect indeed knows infinitely more propositions than we do, since it knows all. Yet with regard to those few which the human intellect does understand, I believe that its knowledge equals the Divine in objective certainty – for here it succeeds in understanding necessity, than which there can be no greater certainty. (D 103)

By specifying geometry and arithmetic alone, Galileo deliberately excluded physics and astronomy, which involve events of Nature. Those, as he had said, elude perfect understanding: but the necessity we experience in mathematical proof gives us a taste of the sureness that Nature displays in her operations. For the most part, at any rate, the means we must employ to link mathematics with Nature are neither certain, nor capable of unrestricted and unqualified application:

No firm science can be given of such things as weight, speed, and shape [of bodies in motion], which are variable in infinitely many ways. Hence to deal with such matters scientifically, it is necessary to abstract from them. We must find and demonstrate conclusions abstracted from the impediments [of material] in order to make use of them in practice under those limitations that experience will teach us. (TNS 225)

In that way Galileo conceived of the union of practical experience with abstract science which was characteristic of the Scientific Revolution. Concerning the future of philosophy, he wrote:

Philosophy itself cannot but benefit from our disputations, for if our conceptions prove true, new achievements will have been made; if false, their refutation will further confirm the original doctrines. So save your concern for certain philosophers; come to their aid and defend them. As to science, it can only improve. (D 37–8)

The above passage, in Galileo's *Dialogue*, was addressed to an Aristotelian who feared that Galileo's science would bring philosophy tumbling down. Galileo was under no illusion that philosophers would even try to understand his science, let alone give up any of their views:

There is no danger that so large a multitude of great, subtle, and wise philosophers will allow themselves to be overcome by one or two who bluster a bit. Rather, without even directing their pens against those, by means of silence alone, they place them in universal scorn and derision. It is vanity to imagine that one can introduce a new philosophy by refuting one author or another. It is necessary first to teach the reform of the human mind, and render it capable of distinguishing truth from falsehood, which only God can do. (D 57)

Galileo shared with Bacon and Descartes the dream of a new philosophy that would displace the verbal exercises of Aristotelianism, but unlike them he did not attempt to set one forth. That seemed to him to lie far in the future, after a great deal more was known about the physical universe as the result of joining practical experience with reason in what I have called useful science. This had begun to appear in the sixteenth century, outside the universities, since it could contribute nothing to natural philosophy as codified in academic instruction. Useful science differed from practical knowledge by systematically organising it for the first time. Natural philosophy was already tightly organised and complete; any change in or addition to it would alter metaphysics, derived by Aristotle from his natural philosophy, and would thereby affect the rest of philosophy. The completely integrated and unified character of philosophy constituted its strength, and at the same time made it necessary for science to proceed independently of it in order to advance at all.

Science remained a natural monopoly of the universities as long as the only permanent records of knowledge were manuscripts. The first printed books were usually expensive, came out in small editions, dealt mainly with topics of interest to scholars and theologians, and (like manuscripts) tended to accumulate only in centres of learning. This situation changed around 1500. By then there were printers in many cities, with investments that made it desirable to keep their presses busy. Led by Aldus Manutius at Venice, they began to issue cheaply printed books appealing to wider audiences, and to solicit books of interest to the public from new authors. Literacy grew rapidly during the Protestant movements, which generated many tracts and pamphlets that were countered by Catholic propaganda, both sides expanding education so that these would be read. Encouraged by printers, authors multiplied, some writing to educate the public and others to impart new knowledge and practical information. Useful science thus spread far from centres of learning.

Universities perhaps benefited less than any other segment of society from the outpouring of cheap books. They had been flourishing for centuries without multiple copies of texts and continued to do so, through lectures and debates. University science was, if anything, less advanced than it had been in the fourteenth century. Commentaries on Aristotle remained the principal texts. The traditional task of the professor was not innovation but selection and preservation of accepted material and its transmission to students.

With the exception of medicine, the most important advances in science during the sixteenth century originated outside the universities. The new astronomies of Copernicus and Tycho Brahe, the mathematics of Tartaglia and Stevin, the mechanics of Guidobaldo del Monte and the physics of G. B.

Benedetti are examples. Such developments, moreover, rarely passed into university instruction; only two or three professors in sixteenth-century Europe, and none in Italy, were interested enough in Copernicanism to teach it. The questions of physics that professors of philosophy debated had nothing to do with the rise of useful science, but had originated in the Middle Ages.

The extent to which Galileo was indebted to medieval science and philosophy has been a matter of dispute. Until the present century it was generally supposed among historians of science that the Middle Ages constituted a sterile period between the Ptolemaic era and the time of Copernicus so far as astronomy is concerned, while in mathematical physics nothing of importance was recognised between Archimedes and Galileo, an even longer time. Those views were seen to be very wide of the truth when Pierre Duhem published his researches among medieval manuscripts, mainly around the time of the First World War. Duhem's discoveries were so numerous that he came to regard science as having evolved continuously from classical antiquity to the present, with occasional periods of strikingly rapid advance, and he expressed the challenging view that if there had ever been a scientific revolution it took place not in the seventeenth but in the fourteenth century. Duhem's conclusions have been modified by subsequent researches, though in many respects they are still viable.

Duhem's work brought up two questions in relation to Galileo's science. First, the specific one whether his important work on the fall of heavy bodies had been derived from direct investigation of nature, in the spirit of modern science, or grew out of medieval impetus theory and the simultaneous (but independent) 'mean-speed' analysis of uniformly accelerated motion (see below). The second, wider question was whether Galileo's science was grounded in philosophy and did not represent a challenge to it, or was a rival approach to the understanding of nature – as it had always seemed to be from his derogatory references to philosophers in his two last and most significant books. Since medieval science certainly was a branch of Aristotelian philosophy, Duhem's continuity thesis required the portrayal of Galileo's physics as rooted in medieval works philosophically, not in new direct investigations of natural phenomena. On the other hand, the violent attacks launched by Peripatetics against Galileo's science made it difficult to regard him as an Aristotelian natural philosopher.

Another approach to Galileo's science was initiated in 1939 by Alexandre Koyré, who saw it as a Platonist reaction to the traditional Aristotelianism of the universities. Galileo's emphasis on mathematical physics was rooted, for Koyré, in Plato's doctrine that the only world worthy of a philosopher's study was inaccessible to the senses and could be grasped through mathematics alone. Medieval mathematical investigations of motion in the abstract, though conducted by Aristotelians, had paved the way, but the Platonism of Galileo constituted a true revolution in science, according to Koyré. In his opinion Galileo's alleged experiments were purely imaginary, and all his studies of motion could be accounted for by mathematical reasoning in the style of Archimedes. Such conclusions gained the support of the majority of historians of science.

Medieval continuity in the literal sense may be rejected on the evidence of Galileo's own manuscripts. No trace of 'mean-speed' in accelerated motion, essential to medieval mathematicians, is to be found either in Galileo's working papers on motion or in his published books. Medieval mean-speed analysis related accelerated motion to uniform motion through the speed at the middle instant of time. There is a middle instant only for a completed finite motion. Galileo reasoned about acceleration for open-ended motions with mathematically continuously changing speed. As to the medieval idea of 'impetus', that quickly gave way in Galileo's writings to simple conservation of speed as such. Previously unpublished notes of Galileo's have recently also disclosed records of experimental measurements, invalidating the conclusion of Koyré. Finally, mathematical concepts and procedures quite different from those of medieval natural philosophers were adopted by Galileo in reaching his basic law of free fall of heavy bodies.

In a third approach, historians of scientific method, led by J. H. Randall, Jr., undertook special studies of Paduan Aristotelianism in the sixteenth century, and produced a new Aristotelian interpretation of Galileo's science. During the

Renaissance, professorial discussions of method and of the certainty of mathematics led to a kind of enlightened Aristotelianism at the University of Padua. Galileo adopted some of its terminology, and according to these scholars his method in science was borrowed from that source. The problem in this case is that Giacomo Zabarella, a leader of the new Aristotelianism at Padua who died only shortly before Galileo began to teach there, was outspoken against the use of mathematics in science, though he favoured direct appeal to experience. His successor was Cesare Cremonini, who steadfastly opposed Galileo on every scientific issue while both were teaching at Padua. The two men were personally friendly, but it is clear that the final official flowering of Paduan Aristotelianism was in opposition to Galileo's advocacy of careful measurement and use of mathematical proportionality in place of physical principles obtained by induction and the quest for causes through syllogistic logic.

The search for some philosophy on which Galileo might plausibly have founded his conception of science has led others to the atomism of Democritus, but only by total misunderstanding of Galileo's mathematical analysis of continuous magnitude. Still others make of him a 'conciliator', as those used to be called who reconciled Plato with Aristotle, or else a philosophical eclectic who (like Giordano Bruno) took bits and pieces of conflicting philosophies to suit themselves. Some see Galileo as a precursor of the philosophical empiricism of John Locke; others, of the positivism of Auguste Comte. The fact is, as pointed out by Alistair Crombie years ago, that hardly a philosophy can be named that does not find something in Galileo's writings to give it aid and comfort. It is therefore quite easy to associate Galileo's science with this or that philosophical system, for whatever that is worth.

Since university physics remained Aristotelian natural philosophy while Galileo was a student, it is understandable that during his first years as a professor of mathematics, he attempted little more than the improvement of conventional treatments of motion – and even for doing that he felt the antagonism of professors of philosophy who taught physics. In private teaching, however, he encountered and became interested in practical problems involving mathematics. His solutions of these led him to useful science and also to recognition of the ingenuity of practitioners who, often without much education, solved physical problems. Ultimately he published most of his books in Italian, writing to a friend in 1612:

> What inspires me to do this is my seeing how students in the universities, sent indiscriminately to become doctors, philosophers, etc., apply themselves in many cases to such professions when unsuited to them, while others who would be apt are occupied with family cares or with other pursuits remote from literature. Though well provided with horse sense, as Ruzzante would say, such men, being unable to read things written in Latin, become convinced that these wretched pamphlets containing the latest discoveries of logic and philosophy must remain forever over their heads. Now, I want them to see that just as Nature has given them, as well as philosophers, eyes to see her works, so she has also given them brains capable of grasping and understanding them. (GW 187)

To the question of who could replace Aristotle as a guide in philosophy, Galileo replied:

> We need guides in forests and unknown lands, but on plains and in open places only the blind need guides. It is better for such people to stay at home, but anyone with eyes in his head and with his wits about him could serve as a guide for them. (D 112)

The idea of addressing ordinary intelligent people with a view to opening their eyes to Nature's works while providing them with no philosophical guidance at all did not become generally popular among scientists until the nineteenth century, when T. H. Huxley called science 'organised common sense'. To more recent scholars it has seemed an utter anachronism to credit Galileo with a similar view, no matter what he may have said. Some, convinced that science without philosophy must always have been impossible, finding that Galileo neglected to expound a philosophy, have seen it as the task of historians to discover or create one for him. But it was precisely philosophy to which Galileo referred when he spoke of forests and unknown lands, and again when he wrote:

> Such profound contemplations belong to doctrines much higher than ours, and we [as scientists] must be content to remain the less worthy artificers who discover and extract from quarries that marble from which, later, able sculptors cause to appear marvellous figures

that lay hidden beneath those rough and formless exteriors. (TNS 182–3)

Galileo did not believe that science, as a method of demonstration and reasoning capable of human pursuit, would ever answer all questions of interest to humanity, or even very many of them. How he arrived at that view is the story of his life and work.

2 Galileo's early years

Galileo Galilei was born at Pisa on 15 February 1564. His father, Vincenzio Galilei, was a musician whose originality and polemic talents fomented a revolution uniting practice and theory in music much as Galileo was to unite them in science. Galileo's mother, Giulia Ammannati, is known but slightly from a few letters that give us an unflattering picture of her.

Galileo was the oldest of seven children. The family remained at Pisa until he was about ten years old and then moved to Florence. After some schooling there he was sent to the ancient Camaldolese monastery at Vallombroso, where he was so attracted by the quiet and studious life that he entered the order as a novice. His father, however, wished him to study medicine and took him back to Florence, where Galileo continued his studies with the Camaldolese monks, though no longer as a prospective member of the order, until his matriculation at the University of Pisa in 1581.

About the same time there was created at Florence an informal academy called the Camerata, active in the literary, artistic, and especially the musical life of the city, which attracted Galileo's father. After having studied music theory at Venice under Gioseffo Zarlino some years earlier, Vincenzio became interested in the revival of classical Greek forms of music as an antidote to the over-ornate vocal polyphony of his time, and in problems of instrumental music as related to a single voice. These pursuits, taken up at the Camerata, led to a sharp controversy between Vincenzio and Zarlino over music theory, which had become abstractly mathematical in ways that retarded innovation. There could hardly be a better example of the Aristotelian division between *episteme* and *techne* than traditional musical science, with its sterile debates about theory, and the rapidly changing musical practice that led to the birth of opera and the development of harmonic modulation soon after Vincenzio's fight against pure theory.

Galileo's early years at the University of Pisa earned him a reputation for contradicting his professors. In a note he wrote many years later he told of his doubt, on first beginning to study Aristotle's natural philosophy, that bodies really fell with speeds proportional to their sizes. He had seen hailstones of very different sizes striking the earth together, which common sense would assume to have started their fall together, from about the same height. According to Aristotle's conception the larger stones should have arrived first, and the smaller ones last. That was not what was seen. The demand that science conform to actual observation is now taken for granted, but that was not a main consideration in Aristotelian natural philosophy, which was content to explain how things should happen in agreement with qualitative causal principles.

In 1583 Galileo sat in on some lectures on Euclid's geometry, not at the university but by a practical mathematician in the service of the Grand Duke of Tuscany. These inspired him to start the study of Euclid's *Elements* on his own. The court mathematician, Ostilio Ricci, quickly recognised Galileo's talent from the questions he brought to him. Accordingly he asked Vincenzio to let Galileo concentrate on mathematics, but the father insisted that he first complete his medical course. Galileo nevertheless neglected that, studied mathematics and philosophy, and left the university in 1585 without a degree.

From about this time we have the earliest surviving manuscript written by Galileo. It consists of discussions, probably designed to be used for lectures, on many of the questions in physics and cosmology which occupied professors of natural philosophy in the universities of that time, treated in the standard conventional style. There is hardly any trace of originality, and none of that emphasis on mathematics which permeated Galileo's later compositions. The Copernican astronomy was mentioned, but decisively rejected. Why Galileo should have troubled to compose so long and standard a work is puzzling unless he aspired to obtain a teaching position and for that purpose he needed to prepare lectures for his own use. These have been shown to be patterned on lectures and books by eminent Jesuit professors of the late sixteenth century. Though as a student Galileo had questioned some of Aristotle's conclusions in physics, it is evident that he as yet had no quarrel with the accepted principles of natural philosophy.

The pursuit of physics by hair-splitting quibbles and scholastic logic applied to Aristotle's texts, as in this first long manuscript of Galileo's, was more an elaborate verbal game than an investigation of Nature. Years later Galileo was to write, when replying in *The Assayer* to a scholarly Jesuit opponent in a controversy over comets:

> Here Sarsi gets up in arms and in a long series of attacks he does his best to show me a very poor logician for my having called a certain enlargement 'infinite'. At my age these altercations simply make me sick, though I myself used to plunge into them with delight during my youth. . . . Sarsi has indeed a large field here for showing himself a better logician than all the authors in the world, among whom I assure him that he will find the word 'infinite' chosen nine times out of ten in preference to 'extremely large'. (D&O 241)

For a few years after leaving the university, Galileo offered private instruction in mathematics at Florence and Siena. His first original scientific treatise was written in 1586, on the hydrostatic balance; it showed a mixture of theoretical and practical interests, the former being taken from Archimedes. About the same time he began writing a treatise on motion which, after revisions and additions made during the next four or five years, afforded the basis from which he later embarked on his most important contributions to physics.

Meanwhile Galileo's father, by experiments on the lengths and tensions of musical strings, discovered a mathematical law that contradicted the fundamental assumption of traditional music theory. Galileo probably witnessed these experiments and kept them in mind later when he sought a rule for the changing speeds of falling bodies. His father's writings also resembled in various ways those of Galileo in his scientific controversies; thus Vincenzio had written in his *Dialogue on Ancient and Modern Music*:

> It appears to me that they who in proof of anything rely simply on the weight of authority, without adducing any argument in support of it, act very absurdly. I, on the contrary, wish to be allowed to raise questions freely and to answer without any adulation [of authorities,] as becomes those who are truly in search of the truth. (PLG 2)

Galileo's mathematical abilities were already becoming recognised even among leaders in Florentine literary circles. In

1588 he was invited to address the Florentine Academy on the location, size, and arrangement of hell as described in Dante's *Inferno.* We now think of the *Divine Comedy* as poetry, not as science, but Dante had skilfully incorporated in it the accepted science of his time. His conception of the infernal regions had been actively debated throughout the sixteenth century, two opposed views having been set forth by commentators on Dante's text. On geographical and mathematical grounds, Galileo supported the earlier of those views. The influential head of the literary academy subsequently aided Galileo in obtaining both his successive professorships of mathematics, first at Pisa and then at Padua.

About the end of 1587 Galileo had discovered an ingenious and practical approach to determination of the centres of gravity of certain solids, an advance beyond Archimedes which earned him his first recognition abroad. On the strength of this discovery Galileo applied in 1588 for a vacant chair of mathematics at the University of Bologna. The position was awarded to G. A. Magini, a Paduan astronomer who already had some published books to his credit, but Galileo's discovery aroused the interest of the Marquis Guidobaldo del Monte, author of an important book on mechanics who from this time to his death in 1607 was Galileo's friend and patron. The same discovery had opened Galileo's acquaintance with Christopher Clavius, mathematician and astronomer at the Jesuit college in Rome, during Galileo's first visit to that city late in 1587.

The chair of mathematics at the University of Pisa was given to Galileo in 1589. It was a poorly paid position, the study of mathematics being regarded as of minor importance at Pisa, but it established Galileo as a professor who could reasonably aspire to the more distinguished post at Padua. His two patrons began at once to work toward that end.

At the same time that Galileo began teaching at Pisa, a distinguished Dante scholar named Jacopo Mazzoni was appointed to teach philosophy there and became his friend and counsellor. Mazzoni later published a book comparing the philosophies of Plato and Aristotle; commenting on it in a letter, Galileo recalled their disagreements during his days at Pisa. In their philosophical discussions they were joined by Girolamo Mercuriale, professor of medicine, whose book on gymnastics and health pioneered that field. It is evident that much of Galileo's time at Pisa was spent in discussions with older colleagues on matters of broader interest than mathematics, though his talents in that field were already such as to impress a visiting Roman mathematician, Luca Valerio.

Galileo at this time still accepted the earth as the centre of the universe and wrote a commentary on Ptolemy's *Almagest* though he was already familiar with the work of Copernicus. These things were mentioned in his Pisan manuscript *De motu*, several chapters of which made direct attacks on Aristotle's physics. The whole treatise may be called pre-scientific, mixing causal ideas taken from Aristotelian philosophy with mathematical ideas taken from Archimedes. At first Galileo expected to reconcile these by removing some errors of Aristotle's, while retaining his basic concept of natural philosophy.

Of special interest in the Pisan *De motu* are Galileo's arguments for equal times of fall for bodies of the same material, regardless of weight, through the same medium. Historians generally have doubted the story about Galileo and the Leaning Tower of Pisa, first told after Galileo's death by a protégé who was not born until long after the incident. According to this story, the demonstration was performed in the presence of Galileo's students and some professors. It is probable that his students, who had been taught Aristotle's rules by their professors of philosophy, would argue against him that weight must affect speed of fall. Galileo's Leaning Tower demonstration would then have been not just to show the students, but to convince the professors that Aristotle's physics must be revised, as he was already arguing.

In a later dispute (1612) with Galileo, a professor of philosophy at Pisa conducted experiments from the Leaning Tower to support Aristotle, observing that bodies of the same material and different weights do not hit the ground *exactly* together. The basic difference between his approach and Galileo's is illustrated in Galileo's last book:

> Aristotle says that a hundred-pound ball falling from a height of a hundred cubits hits the ground before a one-pound ball has fallen one cubit. I say they arrive at the same time. You find, on making the test, that the larger ball beats the smaller one by two inches. Now, behind those two inches you want to hide Aristotle's ninety-

nine cubits and, speaking only of my tiny error, remain silent about his enormous mistake. (TNS 68)

In Aristotle's science every part was linked logically to every other, so it seemed to his followers that nothing he said could be wrong. Galileo remarked:

If Aristotle had been such a man as they imagine, he would have been of intractable mind, obstinate spirit, and barbarous soul – a man of tyrannical will who, regarding all others as silly sheep, wished to have his own decrees preferred over the senses, experience, and Nature itself. But it is the followers of Aristotle who have crowned him with authority, not he who usurped it or appropriated it to himself. (D 110)

I often wonder how it can be that these strict supporters of Aristotle's every word fail to perceive how great a hindrance they are to his credit and reputation, and how the more they desire to increase his authority, the more they actually detract from it. For when I see them being obstinate about sustaining propositions which I personally know to be transparently false, and trying to persuade me that what they are doing is truly philosophical and would be done by Aristotle himself, it much weakens my opinion that he philosophised correctly about other matters, more recondite to me. (D 111)

Galileo's *De motu* was better than its printed rivals and contained things not previously known. It would have been advantageous to him to publish something in his quest for a better position. Yet he withheld *De motu* from the press, probably because his conclusions about speeds on inclined planes did not meet the test of actual experiment, as he candidly admitted. He attributed this to 'material impediments' and added certain theoretical considerations, but in fact it was his neglect of acceleration that rendered his first conclusions very wide of the truth. In any event his withholding of *De motu* from publication was in character, for as he wrote on a later occasion, in his *Letters on Sunspots*:

Even the most trivial error is charged to me as a capital fault by enemies of innovation, making it appear better to remain in error with the herd than to stand alone in reasoning correctly. I may add that I am quite content to be last and to come forth with a correct idea, rather than get ahead of other people and later be compelled to retract what might indeed have been said sooner, but with less consideration. (D&O 90)

As Galileo's three-year appointment at the University of Pisa drew near its end, he had reason to believe that it would not be renewed. Though he had found a few close friends among his colleagues, he had antagonised other professors and had also made a powerful enemy close to the Tuscan court by adversely criticising a scheme to improve the harbour at Livorno. The death of his father in 1591 had also left him responsible for meeting the terms of a generous dowry bestowed on his eldest sister, Virginia. In 1592, on the strength of his teaching at Pisa and with the support of his earlier patrons, Galileo was appointed professor of mathematics at the University of Padua with a salary three times what he had been paid at Pisa.

The University of Padua was renowned throughout Europe for its school of medicine, where Vesalius had taught and where Fabricius of Acquapendente, later the teacher of William Harvey, had recently become professor of anatomy. Padua was hardly less prominent in philosophy. Giacomo Zabarella, the leading Renaissance exponent of Aristotelian method in natural philosophy, died there in 1589 and was succeeded by Cesare Cremonini, whose determination to defend Aristotle's every word made him the model philosopher in Galileo's mature dialogues. In mathematics, Padua was second only to Bologna among Italian universities; indeed Magini, who had won that chair in competition with Galileo, tried to gain the professorship at Padua in preference to it.

Situated twenty miles from Venice on the mainland, Padua had passed under Venetian sovereignty about a century before. Its university benefited from the enlightened government of Venice, which surpassed all other Italian states in tolerance. There was also an active intellectual community at Padua apart from the university. This centred at the home of G. V. Pinelli, who owned a great collection of manuscripts and books and who frequently received visiting dignitaries and scholars and invited Paduan men of letters to meet them. Galileo was lodged for a time by Pinelli and remained his close friend until his death in 1601. It was probably at his house that Galileo met Fra Paolo Sarpi and Robert Cardinal Bellarmine, both of whom were to play important roles in his career as a scientist. The fact that he was greatly respected by both men, whose own views were in sharp opposition on a momentous issue, is highly significant for any assessment of Galileo's abilities and his personality.

Sarpi, a Servite friar, is best known for his activities as official theologian to the Republic of Venice in 1606. In that year, after long friction between Rome and Venice over matters of papal power in secular affairs, Paul V placed Venice under the interdict on the advice of Bellarmine as his personal theological consultant. Sarpi advised the Venetians to ignore this and to command priests to continue their services or face civil punishment. Jesuits were expelled from Venetian territories; other priests remained, and there was little disruption of daily life. Sarpi and Bellarmine engaged in sharp polemics, while practical victory went to the Venetians; a nearly successful attempt on Sarpi's life was generally blamed on the Jesuits. Prior to these events Sarpi was active as a keen student of philosophy and science, as he continued to be on a reduced scale. The researches of Fabricius of Acquapendente on the valves in the veins, of great importance to Harvey in his discovery of the circulation of the blood, are said to have been inspired by suggestions from Sarpi. It is in his notebooks that the earliest description of Galileo's tide theory is found, and it was to Sarpi that Galileo first wrote of his law of fall in 1604 and of his telescope in 1609.

Bellarmine, a Jesuit, was one of the cardinals of the Inquisition who sentenced Giordano Bruno to death at the stake in 1600 for heresy. It is not uncommon to link the condemnation of Bruno with Copernicanism, but the issues in his case and in Galileo's were totally different. Bruno was truly a Copernican zealot, but if that was connected with his ultimate fate at all, it was connected only indirectly. Bruno dreamed of restoring universal harmony in religion through the adoption of an all-embracing philosophy, which the inquisitioners judged heretical after Bruno had been repeatedly denied his right of appealing all questions of heresy to the pope. To the cardinals who condemned him to death Bruno said: 'You must feel more fear in pronouncing this sentence than do I upon hearing it.' Bellarmine can never have forgotten those words with their clear and just accusation against judges who had denied a defendant his legal right of final appeal. That memory probably influenced his recommendations and actions in the events of 1615–16 which affected Galileo.

Bellarmine made a visit to Padua expressly to meet Pinelli, accompanied by another cardinal, Cesare Baronius. Outside the city they changed into humble clothing and arrived at Pinelli's home not as princes of the Church but as monks. Only after having been received graciously and extended every hospitality did they reveal their identities. The event tells us much about the civility of the circles in which Galileo moved at Padua, but its main interest relates to the likely source of a saying by Baronius cited by Galileo in 1615: 'The Bible tells us how to go to Heaven, not how the heavens go.' He certainly did not find this in the works of the learned Church historian, and probably had heard it in conversations at Pinelli's home.

After Pinelli's death the principal meeting-place of Galileo's literary friends became the home of Antonio Querengo, a canon of the Church and a Latin poet of distinction. A favourite diversion of these friends was the use of rustic Paduan dialect, made popular among intellectuals early in the sixteenth century by a writer who called himself Ruzzante. His speciality was the writing of dialogues and plays ridiculing the sophisticated Arcadian poets who glorified simple country life. Ruzzante showed that as it was – hard, but relieved by common sense and vulgar humour. Galileo took particular delight in this down-to-earth literature.

The University of Padua attracted many young foreign noblemen destined for military careers. Mainly for their benefit and to augment his salary, Galileo offered private instruction in military architecture, fortification, surveying, mechanics, and related subjects not included in the university curriculum. In 1593 he wrote outlines of courses in mechanics and fortification, adding others later. There is no evidence of his having shown any special interest in astronomy before 1595, when he hit upon a mechanical explanation of the tides for which were required the two circular motions of the earth assumed by Copernicus. This appears to have marked the beginning of his preference for the new astronomy.

At that time, before the telescope, the evidence for the Copernican system was not very compelling. The leading astronomer after Copernicus was the Dane Tycho Brahe, who rejected motion of the earth as contradicting both the Bible and ordinary events seen on earth as explained in Aristotelian physics. Tycho's scheme of holding the earth fixed but placing the planets in orbits around the sun had won at least as many adherents as the Copernican astronomy, though most scholars still accepted the Ptolemaic system, which placed all revolutions around the earth. On strictly astronomical grounds the choice mattered little. Philosophically, the Ptolemaic system was easiest to reconcile with Aristotle, though that required belief in solid crystal spheres carrying the planets, against which Tycho had brought firm evidence from comets. For those who accepted that evidence, Tycho's system conveniently avoided a break with Aristotelian physics.

Mazzoni's book comparing Plato and Aristotle, which appeared in 1597, contained a fallacious argument against the Copernican astronomy to which Galileo replied in a long letter. This was his first known expression of preference for Copernicanism. Later the same year a German visitor left with Galileo the first book published by Johann Kepler, which was enthusiastically Copernican. In thanking Kepler, Galileo said that he had long accepted the new astronomy, having explained by it some things that could not be otherwise accounted for, but added that he did not teach it publicly because its numerous foolish opponents made that dangerous. Kepler did not ask what Galileo thought he had explained (though he correctly guessed that it must be the tides), but he requested Galileo to make certain astronomical observations if he had accurate instruments, the object being to verify annual motion of the earth by stellar parallax. Galileo did not make the attempt, having probably no special instruments and certainly no hope of finding astronomical evidence that had escaped the best observers. The first known astronomical observations made by Galileo were carried out in 1604 for different reasons.

For his military students Galileo had written a treatise on sighting and triangulation, which was followed in 1597 by his invention of a mechanical calculating device he called the 'geometric and military compass'. Originally conceived for the solution of a practical artillery problem, it was then improved until it could yield rapid approximate solution of any practical mathematical problem then likely to arise. In 1599 he employed a craftsman to make these instruments for sale and began offering each year a course of instruction in their uses.

About the same time Galileo, who never married, formed a liaison with a Venetian woman, Marina Gamba, who bore him daughters in 1600 and 1602 and a son in 1606. She remained at Padua when Galileo returned to Florence in 1610. During these years Galileo's financial situation was made difficult by his promise of a generous dowry for his younger sister, Livia, who married in 1601. His brother, Michelangelo, had moved to Poland with funds borrowed from Galileo which he did not repay, nor did he pay his half of the dowry, of which the down payment alone was double Galileo's salary. Later Michelangelo moved to Germany where he married, obtaining from Galileo funds for that expense, and eventually he sent first his son and then the rest of his family to live with Galileo in Florence.

Galileo increased his private teaching, obtained advances against his salary, and borrowed money from Giovanfrancesco Sagredo, a Venetian gentleman who had studied with him and became one of his closest friends. He was a talented amateur of science, active also in governmental affairs. In particular Sagredo much advanced thermometry, starting from a thermoscope invented by Galileo and applied to medicine by their friend Santorre Santorio, then a physician at Venice and later professor at the University of Padua. Santorre greatly contributed to experimental medicine, Galileo having been one of his subjects in his studies of human metabolism.

Until about 1602 Galileo had been mainly occupied at Padua with practical rather than with theoretical investigations. During that year he completely revised his treatise on mechanics and resumed his earlier studies of natural motion, discovering his first two correct theorems concerning motions along inclined planes. It is of interest that both theorems were derived from Galileo's old incorrect assumptions in *De motu*. We tend to suppose that correct conclusions cannot be obtained by valid reasoning from false assumptions, though only the reverse is the case – that is, incorrect conclusions cannot be logically deduced from true assumptions. At the beginning of modern science it was usually an incorrect assumption that first led to some new truths, which were then placed on firmer foundations after the original supposition was found to yield false as well as true conclusions.

Late in 1602 Galileo wrote to Guidobaldo del Monte about his findings and added the conjecture that from any point on a vertical circle, a body would reach the lowest point in the same time, which is only approximately correct. Guidobaldo tried experiments using the rim of a large winnowing sieve and found it not true, to which Galileo replied that unevenness and friction might interfere and could be eliminated by substituting a long pendulum. Clearly he was concerned with actual experiments and had become good at designing them, though in the same letter he remarked that absolute agreement with mathematical exactitude was not to be expected. His work with pendulums at this time suggested to Santorio the invention of the 'pulsilogium' for use in medical diagnosis, often wrongly credited to Galileo in his student days.

The pendulum was destined to play so large a part in Galileo's scientific work that a few words about it will be in order. It seems curious that its property of swinging in equal times had not been applied long before to time-keeping and used in scientific investigations. A few decades ago it was said that an Arab astronomer in the Middle Ages had represented the pendulum in his book, but this was found to be mistaken. Clocks in Galileo's day were regulated by a horizontal rod driven back and forth through a small angle by a weight attached to a cord, much as pendulums were later kept swinging. The speed of such clocks could be adjusted by moving little weights hung from the rod, but they were never very reliable. Galileo devised an astronomical timer using a pendulum, but the true pendulum clock was invented after his death by Christian Huygens.

What first struck Galileo about a pendulum was not just that it swung back and forth in equal times, but that the time of the swing remained the same whether the arc through which it swung was large or small. That is not exactly true, but it is nearly so, and it seemed paradoxical that the pendulum could adjust its speed so that it took the same time to return through smaller and smaller distances as its motion died down. It was probably the long and heavy pendulum that Galileo used in 1602 which called his attention to the importance of acceleration in downward motion, and to the continuation of motion once acquired – things that soon led him to an entirely new basis for his science of motion which replaced his earlier causal reasoning.

During 1603 Galileo solved several problems of motion on inclined planes and began to study acceleration. The assumption that had been made ever since the fourteenth century was that little successive spurts of speed took place, each speed being uniform while it lasted and greater than the one before. Galileo started out with that idea, but soon had to abandon it. In 1604 he devised a way to measure actual speeds in acceleration. For this purpose he let a ball roll from rest down a very gently sloping plane (less than 2°) and marked its positions after a series of equal times, judged by musical beats of about a half-second. These distances were then measured in units of about one millimeter, from which Galileo obtained the rule that successive speeds in descent follow the odd numbers 1, 3, 5, 7 ..., so that the accumulated distances from rest are as the numbers 1, 4, 9, 16 ..., which gave him the law of falling bodies – that distances from rest are as the squares of the elapsed times.

Knowing this, Galileo progressed rapidly without further experiments, because once mathematics is tied to actual measurements it can be trusted on its own. All the previous debates among philosophers about the 'certainty of mathematics' had been detached from actual measurement, as Galileo was soon to point out in a different context. His new and true theorems, however, did not help him to find at once a correct assumption from which he could prove the law of falling bodies itself. In October 1604 he wrote to Sarpi that he had found a proof, but Galileo was mistaken. He reasoned that since when the same weight is dropped through different distances, its impact is proportional to the distance and only its speed has changed, then its speeds in fall must be proportional to the distances fallen. In fact, it is the *square* of the speed that produces the impact effect, and it took Galileo more than three years to realise his original mistake.

The new basis for Galileo's science of motion was careful measurement, through which he began to replace the ancient search for causes with the modern search for physical laws. By the time he came to write his books on science, Galileo had come to take it for granted that measurement was the key to sound physics and he did not make a special point of saying so in books. That may seem odd, but it is not unusual in the history of science that new techniques and procedures without which discoveries would not have been made are not seen at the time as important for their own sake, the discoveries appearing to be much more interesting and significant. Galileo did sometimes describe his procedures of measurement in his books, and if he neglected to expound actual measurement as fundamental to his discoveries, there are reasons for this, which have a bearing on the independence of his physics from past natural philosophy.

First, the entire science of astronomy had depended on careful measurement from the very beginning. It might even be said that astronomy is nothing else but careful measurement of angles and times, and the quest for laws relating those measurements systematically. Astronomical science was always taught in the universities by the professors of mathematics, while an entirely different discipline, taught by professors of philosophy, was based on Aristotle's book *On the heavens.* This discipline, which may be called cosmology, was the true science of the heavens in the sense of *episteme*, while astronomy was in effect not true science at all, but *techne*, in the eyes of proper philosophers. Measurement played no part at all in Aristotelian cosmology, which left these mundane occupations to mere practitioners. This attitude had taken shape in antiquity, when Hipparchus (*c.*150 BC) laid the basis for the later Ptolemaic astronomy in which measurement and calculation alone mattered and all causal or physical explanation was left to the philosophers, who were trained for it. Accordingly there existed an old tradition in science, if we count astronomy as science and not just *techne*, and Galileo saw his physics as no more than the application of astronomical methods to the investigation of motion.

In retrospect, it seems remarkable that that had not been done long before. It had been done in mechanics, but mechanics was not regarded as a part of physics before the seventeenth century, and was confined largely to statics. Motion is much harder to measure than the weights and distances which concern statics, besides which the *causes* of motion, which alone interested physicists, were not obtainable through measurements. Thus it came about that medieval physicists dealt with measure only in the abstract and did not even attempt careful measurements of actual motions. Galileo did, but he made no special point in his books of the measurements that had enabled him to make discoveries about motion, since to him they were just the natural extension of a procedure introduced long ago by astronomers.

The further point to be made, though to go into its full implications for science would be out of place here, is that measurement requires units of some kind, and there is a limit to the accuracy with which actual measurements can be made in terms of counted units and fractions of units. For his measurements of distances covered in actual motion, Galileo's unit was about a millimeter, and he recorded no smaller fractions of that unit than one-half, which is the practicable limit when using a ruler and the naked eye. No matter how we improve techniques of measurement, a practicable limit of accuracy always exists, and people who make actual measurements soon become aware of this. Galileo's father had shown how the advancement of musical practice was hampered by theories assuming impossibly exact measurement; Galileo saw, when he began actual measurements of distances and times, that discrepancies from pure mathematical theory were built into the very processes themselves. Hence he did not insist on the kind of perfection that philosophers had always demanded, and Galileo's science differed from previous natural philosophy because it was based on reasonable agreement with observation rather than on the mind of God or ideals inaccessible to experience, whether mathematical (as with Plato) or verbal (as with Aristotle).

We may call such science slipshod, or we may call it utilitarian; we may despise or admire it; we may prefer natural philosophy to it; but the fact remains that the source of Galileo's discoveries about motion was careful measurement. He took his clue from astronomy, not from the abstract measures of medieval physicists, or from principles of Paduan Aristotelianism, or from Renaissance debates about the source and nature of mathematical certainty. What concerned Galileo was not inside mathematics, nor inside physics, but in relations between the two. His problems in the study of motion from 1602 to 1608 were semantic and mathematical; previous philosophies shed no light on their solutions.

In saying this, I allude to Galileo's analysis of the natural descent of heavy bodies, his chief contribution to mathematical physics. Only quite recently, by study of his working papers, has it become possible to say just what problems confronted Galileo and how he solved them. Few readers will be concerned with the technicalities behind that work as outlined in the next chapter, but because Galileo's procedures have not been taken into account in the usual summaries of his physics, I shall touch on the basic problems that lay behind them.

No one had ever defined 'velocity', even medieval natural philosophers who attacked it mathematically, because Aristotle had already defined 'equal velocity' and 'greater velocity', and that remained sufficient until mathematically continuous change had to be taken into account. Uniform velocity had been analysed by Archimedes, and uniform change of velocity by medieval writers; but uniform change need not be mathematically continuous change. When we count, the numbers change uniformly by jumps of one, but that is not continuous change. Medieval 'degrees of speed' were so counted. Galileo found himself confronted with the continuous change of speed in fall, and to deal with that he had to define 'velocity' consistently with actual physical measurements. That was not easy, because the definition most suitable to impact phenomena would fit speed as measured by the effect of striking, but would not fit the speeds gained during free fall. Thus was created a semantic problem unsuspected by natural philosophers who had not made actual measurements.

A mathematical problem had also to be solved, since 'instantaneous velocity' introduces paradoxes of the infinite unless handled with great mathematical skill. Aristotle's treatment of Zeno's paradoxes of motion had to be extended in a manner useful for rigorous quantitative analysis of continuous change. Galileo achieved that, but only with difficulty and over a period of years. Mathematical techniques had to be devised which, far from having been inspired by philosophical reflections, were opposed by philosophers long after Galileo died.

The key to Galileo's mathematical physics was his application of a theory of proportionality to actual measurements that had been made as accurately as possible by the means at Galileo's disposal. The theory of proportionality he used, though set forth in Euclid's *Elements*, Book V, differed from that used by medieval writers, who had had a defective version of that book. Nor was Galileo's guide any metaphysical belief about nature; rather, it was an epistemological conviction about reliable knowledge. Its counterpart in antiquity was not Plato's philosophy, but Ptolemy's astronomy, which depended on actual measurements, while the former sought eternal truth beyond all possible measurement. Measurement belongs to science; eternal truth belongs to faith, whether philosophical or theological.

3 Conflicts with philosophers

In October 1604, when Galileo was writing to Sarpi about his law of falling bodies, a supernova appeared in the evening sky. Galileo was told about it a few days after a medical student named Baldessar Capra and his mathematics tutor Simon Mayr had observed and confirmed it. A new star had been seen in 1572 and had been proved by Tycho Brahe to be among the fixed stars. According to Aristotle's fundamental principles, no change could ever take place in the heavens, because everything in them was made of a perfect and unalterable substance called the 'quintessence'. Change occurred only in the 'elemental' materials of earth, water, air, and fire. Natural philosophers accordingly taught that comets were not astronomical events, but meteorological phenomena situated in the elemental sphere beneath the moon. New stars could be explained as some kind of tailless and motionless comets, but not as bodies actually in the heavens.

Galileo wrote to astronomers in other cities and compared their observations with his own. Like Tycho's star, this new star exhibited no detectable parallax; no matter where it was observed from, it was seen in the same place with respect to nearby fixed stars. That cannot happen for things as close as the moon. Since people were always excited by unusual appearances in the sky, Galileo gave three public lectures on the new star, explaining how astronomical observations and careful measurements of angles showed that it must be located in the heavens. Aristotle had been simply mistaken.

As the ranking professor of philosophy at Padua, Cesare Cremonini sprang to the defence of Aristotle. It is hard now to realise what a fundamental blow to all natural philosophy it would be if a mere mathematician could prove actual change in the heavens. Cremonini and Galileo were good personal friends and had doubtless debated philosophy and science on many occasions, but this was no friendly discussion: it was a public feud. Cremonini's arguments against Galileo were endorsed in a booklet published at Padua early in 1605, ostensibly by one

Antonio Lorenzini and without naming names. Galileo, who recognised parts of the book as written by Cremonini himself, replied by publishing (over an assumed name) a little dialogue between two peasants, written in rustic Paduan dialect, in which he made a peasant reason better than the celebrated professor.

Cremonini's position was that ordinary rules of measurement on earth did not apply to vast distances. To reason correctly about bodies in the sky, one had to use the Aristotelian principle that distinguished celestial from elemental material. Galileo's peasant spokesman asked what philosophers knew about measuring anything. It was the mathematicians, he said, who had to be trusted in measurements, and they did not care whether the thing seen was made of quintessence or polenta, because that could not change its distance. Shaped by his discovery of the law of fall, Galileo's science from this time on assumed that rules which apply to actual measurements whenever tested will equally apply to situations in which independent test is not possible.

Astronomy had depended on careful measurement for centuries, but not until Galileo's time was that basis extended to physics. Except in optics, there is hardly an instance of the discovery of a mathematical law before Vincenzio Galilei's work on musical strings. The fact that Galileo's law of fall had been discovered by measurement shortly before he became involved in a dispute with a professor of philosophy over astronomical measurement is probably significant to the consolidation of Galileo's science as quantitative and his discarding of Aristotelian 'qualities'. Until recently it was not demonstrable that Galileo had used experimental measurements in his work, and it seemed more sophisticated to believe that his emphasis on mathematics in science represented only a philosophical dogma.

From 1605 on, observation and experiment became for Galileo the solid foundation of science. When possible, he made measurements, and those afforded the only certainty that he ascribed to his conclusions in both astronomy and physics. Aristotle had said that mathematical precision is not to be expected where matter is involved, and at least until 1602, Galileo agreed. Later on he had this very point raised by the

Aristotelian spokesman in the *Dialogue* so that he could reply: 'True, but where it is found, why not make use of it?'

The summer of 1605 was spent by Galileo at Florence as tutor in mathematics to the young prince Cosimo de' Medici. Already well established in the regard of that ruling family, he asked their assistance in securing reappointment to his professorship at Padua, which he knew to be in danger for the first time as a result of his conflict with the philosophers over the new star. He also hinted that the post of court mathematician (left vacant by the death of Ostilio Ricci in 1603) would appeal to him. The practical instruction in mathematics he gave to Cosimo was based on use of his calculating instrument, concerning which he promised to publish a book dedicated to the young prince. The Tuscan ambassador at Venice did assist in his negotiations for reappointment and increase of salary, though at this time there was no response to his suggestion of employment at court.

Galileo's book on the calculating instrument was printed in 1606, in Italian for the benefit of engineers and military men. Early in 1607 Baldessar Capra published a Latin plagiarism of it and hinted that others who wrote of the instrument had stolen it from him. Capra had not even begun the study of mathematics until 1602, under the visiting German tutor Simon Mayr, while Galileo had been making and selling his instrument since 1597. To him it was a serious matter to be accused of having dedicated to the Medici prince something not entirely his own. He obtained affidavits from Sarpi and others who had been early recipients of his sector, one of whom certified that Capra had borrowed from him both the instrument and Galileo's manuscript instructions for its use, some years before. Galileo brought an action against Capra before the governors of the university, showing under cross-examination that the student had not even mastered the contents of the Latin book in question, which had probably been composed mainly by Mayr before he returned to Germany in 1605. Capra was expelled and his book confiscated.

This incident left its mark on Galileo's personality. Up to this time he had been open and free in giving out information and revealing discoveries. The behaviour of Capra, whose father Galileo had befriended in 1604 by recommending him to the

Duke of Mantua, made Galileo secretive about his discoveries and sceptical about the good faith of possible rivals. And despite his publication of a full account of the affair, the mere fact that he had been accused of stealing an invention was used by his antagonists later to cast suspicion on him in the more important matter of his astronomical use of the telescope.

During 1607–8 Galileo put together his previous theorems on motion and added more, seeing finally that speeds in fall are as the square roots of distances rather than as the distances. This gave him a way of testing his previous assumption that horizontal motion would be uniform in the absence of friction. Again the experiment he recorded was remarkably accurate, and again he proceeded from it to many new theorems which he did not submit to separate tests. The most important of these concerned projectile motions, which he discovered to follow parabolic paths as a result of the new experiment. He now began to compose a book on natural motions which was not published until near the end of his life, for reasons presently to appear. One matter of importance to Galileo's physics, in which it differed from Newton's and later physics, will be mentioned here to illustrate his caution about general principles.

Continuance of motion at uniform speed in a straight line ultimately became the cornerstone of Newtonian physics. It is now called 'inertial motion', which Galileo allowed only for heavy bodies moving through relatively short distances near the earth's surface. In his physics, a heavy body must gain or lose speed if it approaches or moves away from the earth's centre; that is, if it falls or rises at all. Over short horizontal distances, as in his 1608 experiment, the body could be considered as remaining at the same distance from the earth's centre, so in his physics the inertial rule was the same as ours for such cases. But Galileo was unwilling to extend that into a universal principle. Indefinite uniform straight motion would imply an infinite universe, and any natural tendency to such motion by heavenly bodies seemed to him inconsistent with the observed order of the cosmos. If any motion in Nature were truly uniform and perpetual, Galileo said, it would have to be circular motion. But he did not assert that any motion in Nature *is* truly uniform; only that relatively brief horizontal motions near the earth may be so considered. That sufficed for terrestrial physics, and Galileo did not speculate about celestial physics as did Kepler.

Nothing better illustrates Galileo's idea of restricting science to things that could be established on 'sensory experiences and necessary demonstrations', in the phrase he adopted when philosophers began to appeal to theologians for support. Newton's extension of the inertial law to all bodies was made when he discovered the law of universal gravitation, confirmed by innumerable astronomical observations. Without universal gravitation, generalisation of the inertial law remained mere speculation, as with Pierre Gassendi and René Descartes soon after the death of Galileo, who had been quite willing to leave such speculation to philosophers and just 'quarry the marble' from which it was carved.

In mid-1609 Galileo was hard at work on his treatise on natural motions when events took place that altered his scientific interests for many years. These began with the invention in Holland of a device that made distant objects appear closer. A patent was applied for to the Dutch government in October 1608, and Sarpi heard about it before the year was out. Either Galileo did not, or he did not believe the rumours, until July 1609. During a visit to Venice in that month he consulted Sarpi about them and was shown a letter from a former pupil of his own, then living in Paris, that confirmed the story. Galileo had again been seeking a salary increase and had just been told there was a little hope. Realising the importance of a spyglass to Venice as a maritime power, he hurried back to Padua to try his hand at making one. There he learned that a foreigner had just passed through Padua with a spyglass which he intended to sell at a high price to the Venetian government.

According to his later account, Galileo reasoned that one of the two lenses must be convex and the other concave, and on fitting such spectacle lenses in a lead tube he found that it worked. That, however, was a mere toy that could magnify only two or three times. Galileo wrote to Venice (doubtless to Sarpi) promising to have a good spyglass very soon. Sarpi, by reason of his reputation in science, had in fact been asked by the Venetian senate to advise them on purchase of the foreigner's instrument. He recommended against it, and late in August Galileo arrived with a telescope about as powerful as our ordinary field-glasses. With this he was able to describe approaching ships two hours before they could even be seen by trained naked-eye observers. He was offered life tenure as professor at nearly double his existing salary when he presented his instrument to the Doge.

There were, however, some misunderstandings. After his acceptance, Galileo learned that no increase was to be made until the expiration of his existing contract, that no further increase was ever to be allowed, and that he was bound to remain teaching at Padua for life. Galileo was a Florentine at heart, and pleasant as Padua had been, it was not home. Nor did he want to oblige himself to teach all his life, wishing more freedom for research and publication. Since he had as yet received no material benefits under the agreement, he felt free to reopen his negotiations for the post of court mathematician at Florence, Cosimo having since become Grand Duke there.

After a hasty visit to Florence, where he exhibited his new instrument to Cosimo, Galileo set to work grinding lenses for a still more powerful one, obtaining glass blanks secretly from Florence so that rivals would not know his design. By the first of December he had a twenty-power telescope, with which he observed the moon on every night that was not cloudy. He correctly interpreted what he saw as proving the existence of mountains and craters where natural philosophers demanded perfect sphericity in the perfect heavens. Early in January 1610 he discovered four satellites revolving about Jupiter, contradicting the idea of natural philosophers that the earth was the centre of all celestial motions. Previously unseen stars in several constellations were charted, and the Milky Way was seen to consist of myriads of stars.

Early in March Galileo published these discoveries in his *Starry Messenger*, dedicated to Grand Duke Cosimo. During the Easter vacation he was invited to visit the court at Pisa, after which the details of his appointment as chief mathematician and philosopher (that is, physicist) to the Tuscan court remained mere formalities.

Galileo's discoveries announced in the *Starry Messenger* produced violent reactions. Among the general literate public they created great excitement, while philosophers and astronomers for the most part declared them optical illusions and ridiculed Galileo or accused him of fraud. A notable exception was Kepler, whose opinion Galileo requested through the Tuscan ambassador at Prague, where Kepler was astronomer to the Holy Roman Emperor. Kepler at once wrote a long 'Discussion with the Starry Messenger' in which he accepted the discoveries as real. A few months later, using a telescope sent by Galileo to the Elector of Cologne, Kepler published confirmation of Jupiter's satellites by his own observations.

Meanwhile a group of persons gathered by Magini at Bologna in April, when Galileo passed through, had been unable to see the satellites even with Galileo present to show them how to use the new instrument. A protégé of Magini's, Martin Horky, sent word of this to Kepler and published a book denouncing Galileo's claims. At Rome, Father Clavius declared his belief that all the new things seen were in the lenses and not in the sky. Several others attacked Galileo's claims in print on astrological and philosophical grounds.

Galileo did not reply, though a friend at Bologna and one of his students at Padua published answers on his behalf. The latter also reported Galileo's use of his telescope to study insects at close range and the imparting of his findings to Cremonini. That philosopher, however, refused ever to look at the sky through the telescope, as did also Giulio Libri, professor of philosophy at Pisa. Galileo delivered three public lectures at Padua and reported to Cosimo's secretary of state:

> The whole university turned out, and I so convinced and satisfied everyone that in the end those very leaders who at first were my sharpest critics and the most stubborn opponents of the things I had written, seeing their case to be desperate and in fact lost, stated publicly that they are not only persuaded but are ready to defend and support my teachings against any philosopher who dares to attack them. (D&O 60)

Support for Galileo began to spread later in 1610, when the Jesuit astronomers at Rome finally obtained a telescope powerful enough to permit confirmation of his discoveries. Clavius, however, entered a dissent concerning the mountainous surface of the moon, which he believed must be an optical illusion.

Mountains on the moon were so objectionable to natural philosophers that lengthy debates were inaugurated over them in Germany, as well as by Jesuits at Mantua, to which Galileo replied patiently and at length.

This dispute over the perfect sphericity of the moon took place mostly in letters and treatises left unpublished, so it has attracted little attention. Yet opposition to lunar mountains throws important light on the prevailing natural philosophy, quite apart from the arguments that the telescope was not to be trusted – taken up again recently by philosophers and historians who rank theory ahead of observation. Some scholars now say that if Galileo had been a good scientist instead of a Copernician zealot, he would have avoided committing himself to the disclosures of his telescope. That is what Clavius and others had said at the time. There is no proof that anything seen through curved glasses exists anywhere except in those lenses, because what is seen disappears when the lenses are taken away. Therefore, it is now said, a complete theory of optics was required before the telescope could be trusted – and Galileo did not have such a theory.

What this modern argument proves is that science lacks philosophical justification, which I believe to be true, but to be no more true of Galileo's science than of our own. What the argument pretends to prove is that early critics of Galileo's discoveries argued on sounder grounds than observation, which I believe to be not true. Even if they had had some complete theory of optics (which they lacked) and could demonstrate the illusory character of all observations, as was done later by Bishop Berkeley, that would not have resulted in our having a better science than Galileo's. It might have resulted in our having no science, but only a philosophy, which might be a blessing, since then we would be free from all doubts. Such was the goal of Galileo's original adversaries, who possessed only a philosophy and wanted things to remain that way.

Aristotelian natural philosophy left no doubt that the moon was perfectly spherical, as were all heavenly bodies. That had nothing to do with optical theory; it hinged on the perfection of the quintessence, which is not a sounder ground for science than is observation. Galileo repeatedly rejected the position that perfection of shape existed at all, except with relation to use, pointing out that spherical bricks would not be perfect for the building of a wall. In the *Dialogue* he said:

> These doctors of philosophy never concede the moon to be less polished than a mirror; they want it to be more so, if that can be imagined, for they deem that only perfect shapes can suit perfect bodies. Hence the sphericity of the heavenly globes must be absolute. Otherwise, if they were to concede me any inequality, even the slightest, I would grasp without scruple for some other, a little greater, for since perfection consists in indivisibles, a hair spoils it as much as a mountain. (D 80)

And the Aristotelian spokesman, when asked why such perfect rotundity was required in celestial bodies, replied:

> Being ingenerable, incorruptible, inalterable, invariant, eternal, etc., implies that celestial bodies are absolutely perfect; and being absolutely perfect entails their having all kinds of perfection. Therefore their shape is also perfect; that is to say, spherical – and absolutely and perfectly spherical, not approximately and irregularly. (D 84)

Several books were published by philosophers in reply to Galileo's *Dialogue* without any complaint that passages such as the above misrepresented the Aristotelian natural philosophy which they supported. The writers were versed in medieval and Renaissance books from which Galileo is now said to have borrowed his conception of science; yet they recognised nothing valid in his arguments. Against his appeals to observation they opposed dogmatic principles. It is true that Galileo offered no complete theory of optics, but he meticulously described experiments with spherical mirrors, plane mirrors, and rough reflecting surfaces. These showed that a perfectly spherical moon struck by the sun would remain invisible to us except for a single bright point, while a moon merely rough, even without mountains, could appear to us much as we see it. Using methods familiar to surveyors, he had measured lunar mountains as high as four miles. Still, his opponents would not concede the slightest irregularity on the moon's surface.

To call those men sounder scientists than Galileo, as many now do, expresses no more than an opinion of what constitutes sound science. Anyone is entitled to prefer philosophy to science. But to assert that Galileo believed planets to move uniformly in perfect circles because he could not shake off the ancient tradition of heavenly perfection is simply false, since he openly derided that tradition in discussing the moon, and remarked elsewhere in the *Dialogue* on irregularities in the motions of both the moon and the sun.

Before leaving Padua Galileo also observed the curious appearance of Saturn, though its rings could not be distinctly seen through his instrument. He moved to Florence in September, having sent his daughters there earlier to be with his mother and leaving his son with Marina Gamba until old enough to leave her care. Soon after his arrival at Florence in September Galileo was able to begin observations of Venus, which had previously been too close to the sun, and discovered its moon-like phases. This showed conclusively that Venus revolved not around the earth but around the sun, destroying the Aristotelian and the Ptolemaic arrangements. It did not establish the Copernican system, since Tycho's astronomy also placed the orbit of Venus around the sun while holding the earth fixed. Galileo, however, correctly regarded the Tychonic system as dynamically absurd, since any power of the sun to carry all the other planets around daily could not leave the earth unmoved.

The phases of Venus were of special interest because they explained one thing that Copernicus himself had noted as very puzzling. If the distance between Venus and the earth varied as much as his system implied, it seemed that the apparent size of that planet should change much more than it did. Now the telescope showed that when farthest from the earth, Venus is entirely illuminated by the sun (like full moon), whereas when visible at its closest, only a thin crescent is illuminated of its much larger apparent disc. Galileo considered it praiseworthy in Copernicus that he had not permitted one unexplained puzzle to stand in his way; had he known the explanation:

> How much less would his sublime intellect be celebrated among the learned! For as I said before, we may see that with reason as his guide he resolutely continued to affirm what sensible experience seemed to contradict. (D 339)

This may appear inconsistent with Galileo's restriction of science, mentioned above, but it is not. Copernicus made no appeal beyond 'sensible experiences and necessary demonstrations', and once this particular puzzle was solved by the former, it was easy to provide the latter. Galileo's point here was that in science it is commendable to reserve judgement with respect to problems not yet solved and to proceed on a preponderance of evidence. It was no scientific solution to introduce arbitrary speculations. Thus in the dispute over mountains on the moon, both Galileo's German opponent and Ludovico delle Colombe, a philosopher at Florence, proposed to defend the moon's smooth spherical surface by saying that it was covered with transparent crystal, beneath which Galileo saw mountains he mistakenly supposed to be on its surface. Colombe sent this idea to Clavius at Rome, where the secretary of Cardinal Joyeuse wrote to get Galileo's reply for that very friendly cardinal. Galileo answered that he would grant his adversaries this crystalline substance if they, with equal courtesty, would then allow him to construct of it mountains ten times as high as those he had actually measured on the moon.

This technique of replying to opponents by drawing consequences compatible with their own assumptions which they themselves could not accept, but could not logically refute, is called argument *ad hominem*. It enables one to destroy a position without even taking any firm position of one's own, and Galileo used it frequently. The logic of such arguments is often missed by his critics today, who believe him to have asserted various things as true which he presented only to deprive specified arguments of the power to prove what they claimed to prove.

The first problem Galileo attacked at Florence was to determine orbits and periods for Jupiter's four satellites. That task was so difficult that Kepler had publicly questioned the possibility of its accomplishment. In December, sending to Prague in anagram form his discovery of the phases of Venus, Galileo mentioned that he was on the track of obtaining the satellite periods as well. (Galileo used anagrams in order to be able to establish dates of discoveries later, if priority claims arose. The letters of words describing a discovery were scrambled, and then sent in a dated letter to a friend. Anagrams were similarly used later by Huygens and by Newton.) Galileo's clue to the satellite periods worked out very well; he had the basic data necessary for predictions by March 1611, and during a visit to Rome in April he began compiling tables of satellite motions.

The first scientific society of lasting significance had been founded at Rome in 1603 by four young men headed by Federico Cesi, who named it the Lincean Academy. Cesi gave a banquet for Galileo at which the word 'telescope' was coined and the guests observed the new discoveries in the heavens. Galileo's move from the university to the court had resulted in his receiving less news of scientific developments; now his election to the Lincean Academy and his subsequent correspondence with its members in Italy and abroad kept him even better informed than before.

At Rome Galileo also renewed his acquaintance with Father Clavius and with Cardinal Bellarmine. The Jesuit astronomers at the Roman college fêted him at a special conference. Several cardinals and other churchmen attended his frequent exhibitions of telescopic discoveries, including sunspots, which Galileo then regarded as mere curiosities. Pope Paul V granted him an audience. There was as yet no sign of theological opposition to Galileo or his discoveries, though Bellarmine wrote to the Inquisition at Venice to know whether he had been involved in proceeedings against Cremonini. Probably this was because Galileo broached to Bellarmine the Copernican implications of his work. Cremonini had nothing to do with that, but he was always in hot water with the Inquisition because he refused to note in his books that certain doctrines of Aristotle had been pronounced heretical, such as the mortality of the soul and the eternity of the universe.

Soon after his return to Florence with many evidences of his hearty reception at Rome, Galileo became involved in a dispute with philosophers over a question of physics. Filippo Salviati, a Florentine patrician who had formed a group that met at his home for intellectual discussions, invited Galileo to join them. In June 1611, they were debating condensation and rarefaction, a fundamental issue between Aristotle and the atomists. Vincenzio di Grazia, a professor of philosophy at Pisa, called ice 'condensed water'. Galileo remarked that it would be better called 'rarefied water', since it floated. Di Grazia replied that that was because of its broad flat shape which could not penetrate the resistance of water against division. Galileo observed that a flat piece of ice held under water and released seemed to penetrate any such resistance, if it existed. He doubted that water resisted a solid's sinking, since the tiniest particles of mud will settle out in time. When it was said that striking a sword flat on water shows its resistance, Galileo agreed that water resists *speed* of motion, but not motion as such – as Aristotle had also said.

Di Grazia mentioned the argument to Colombe, who already had a grudge against Galileo because of an attack on a book he had published about the new star of 1604. He offered to show experimentally that Galileo was wrong and that shape did cause floating, basing this on the floating of thin chips of ebony but not of ebony balls. Meanwhile courtiers had told Cosimo that his mathematician was engaging in disputes that might bring discredit on him, so he advised Galileo to write out his arguments and avoid public quarrels. While Galileo was doing this, a newly appointed professor of philosophy at Pisa was invited to debate the issue with Galileo at a court dinner where two visiting cardinals were present. This resulted in complete vindication of Galileo, whose side was taken by Maffeo Cardinal Barberini, later to become Pope Urban VIII.

Galileo now revised his defensive essay into a constructive treatise on floating bodies, which he wrote at Salviati's villa a few miles west of Florence, while recuperating from illness which he blamed on the city air (but which was probably aggravated by the unaccustomed strains of court intrigues and the growing hostility of philosophical opponents). He adopted a new basis for hydrostatics, using two principles from his mechanics, and explained for the first time how a heavy beam can be floated in very little water. The principle of Archimedes, which he had defended from the beginning and which seemed to be contradicted by the floating of thin chips denser than water, was applied by him to those just as it is applied to a floating tea-kettle. Galileo had observed that such chips float entirely below the surrounding water surface, in little troughs containing enough air to make the overall density of chip and air equal to that of water. To show that flat shape did not assist floating he remarked that little wax cones weighted with metal filings will sink if placed on water base down, but will float if placed point down, the reverse of what would be expected if water resisted division by flat things.

Although a treatise on hydrostatics would not seem likely to have wide appeal, Galileo's book sold out two editions in 1612. Many readers reported that they were convinced after having at first believed Galileo's assumptions paradoxical. Public interest is understandable because of the number, variety, and intrinsic appeal of the experiments Galileo described, requiring no special equipment and being amusing to perform. Four professors of philosophy printed lengthy attacks on Galileo's book, not because Aristotle had said much about floating, but because the basic doctrines of condensation and rarefaction were remotely threatened. In their antagonism towards Galileo they attacked him even where he agreed with Aristotle, as in saying that shape did not affect sinking, but only speed of sinking. Aristotelian natural philosophy had achieved a logical structure such that if any principle were relinquished, everything would have to be altered. Galileo's piecemeal approach to science was designed to prevent any such predicament from ever arising for its followers.

In his book on hydrostatics Galileo remarked that the authority of Archimedes was worth no more than the authority of Aristotle; Archimedes was right, he said, only because his propositions agreed with experiments. In our time a misunderstanding has arisen from the belief that Galileo made mathematics a higher court of appeal than experiment. That is like saying he believed the real positions of Jupiter's satellites were not those he observed, but those he calculated, or that the real positions of a ball along an inclined plane were not the ones he measured. No such statement is found in Galileo, who said:

> When you apply a material sphere to a material plane in the concrete, you apply a sphere which is not perfect to a plane which is not perfect, and you say these do not touch at a single point alone. But I tell you that even in the abstract, an immaterial sphere that is not a perfect sphere can touch an immaterial plane that is not perfectly flat in not one point, but over part of its surface – so what happens here in the concrete happens in the same way in the abstract.
>
> It would indeed be news to me if bookkeeping in abstract numbers

did not correspond to concrete coins of gold and silver or to merchandise. Just as an accountant who wants his calculations to deal with sugar, silk and wool must discount boxes, bales, and packings, so the philosopher-geometer, when he wants to recognise in the concrete those effects which he has proved in the abstract, must deduct the material hindrances; and if he is able to do that, I assure you that material things are in no less agreement than arithmetical computations. The errors, then, reside not in abstractness or concreteness, but in a bookkeeper who does not understand how to balance his books. (D 207–8)

The view Galileo took of the place of mathematics in physics is as different from Plato's view as it is from Aristotle's. Plato regarded the world of pure mathematical ideas as alone worthy of study; if physical objects did not conform to it, so much the worse for them, because they were defective and imperfect anyway. Aristotle considered the procedures of mathematics as alien to physics because mathematicians left matter wholly out of account. Both philosophers were struck by the abstract character of mathematics in contrast with the concrete material world. Galileo, on the contrary, was struck by the utility of mathematics as a tool in the study of physics. Just because calculation did not exactly fit observation was no reason to exalt either, or to abandon either. A poor fit could show that we were leaving something out of account, not that we should leave mathematics or observations out of account.

Once this view took hold, as it did in the seventeenth century, the progress of mathematics was much stimulated by the needs of physicists. Galileo, Descartes, and Newton each devised some mathematical procedures useful to physics for which no need had been felt by pure mathematicians, but which greatly enriched mathematics in their hands. Needless to say, physicists in return received new mathematical tools from them. The old philosophical barriers erected by Plato and Aristotle gave place to new understanding of both physics and mathematics.

4 Conflicts with astronomers and theologians

While Galileo was writing his book on floating in water, a book about sunspots was published pseudonymously by the German Jesuit Christopher Scheiner. Forbidden by his superior to risk discredit to his order, he wrote in the form of letters to Mark Welser of Augsburg, who had previously sent Galileo the German attack against lunar mountains. Welser, a banker to the Jesuits who was soon made a member of the Lincean Academy, printed Scheiner's letters over the name of 'Apelles' and sent them to Galileo for comment, remarking that he did not suppose sunspots were anything new to the Italian.

Galileo received this material on a visit to Florence to place his new book in the hands of a printer. His former pupil, a Benedictine abbot named Benedetto Castelli, had arrived to assist him and was asked to see the book through the press and to make daily observations of sunspots as carefully as possible. Castelli recorded those so accurately that the daily movement of a spot could be measured, enabling Galileo to prove that the spots must be on the sun's surface and that the sun rotated about once a month. Scheiner had concluded that what were called sunspots were really tiny planets revolving around the earth or the sun and obstructing our vision. He wrote additional letters which Welser printed, also answered by Galileo, whose three *Letters on Sunspots* were published at Rome in 1613 under the auspices of the Lincean Academy. The Linceans insisted on adding a preface that Galileo disliked, asserting his priority of discovery.

Galileo had shown sunspots to others while at Rome in 1611, and later the Jesuit mathematician Paul Guldin said that at that time he had sent word of this to Scheiner. Even before Scheiner's book there had been another on sunspots by Johann Fabricius. Yet Scheiner was angered by the priority assertion in the Roman preface, as were many other Jesuits. It started a long and bitter resentment in him that was eventually to have serious consequences for Galileo.

Both Scheiner and Galileo argued many astronomical and

other issues than just sunspots. Galileo took the position that all celestial phenomena should be interpreted in terms of terrestrial analogies, against Aristotle's basic postulate of essential differences. He also asserted that the essences of things cannot be known and that science is concerned only with properties of things and observed events. This amounted to a declaration of independence of science from philosophy.

For the first (and only) time, Galileo came out in print unequivocally in favour of the Copernican astronomy. He had avoided that in the *Starry Messenger*, and even after discovering the phases of Venus he had no astronomical evidence against the Tychonic system. It was only in an appendix to his *Sunspot Letters* that Galileo mentioned what had been for him the clinching fact. This was the discovery of eclipses of Jupiter's satellites and of a simple means of predicting such events. The importance of this as scientific evidence for Copernicanism justifies a brief explanation, especially because Galileo said little of it.

In order to predict positions of the satellites, it was necessary to introduce a correction for the earth's motion – or the sun's motion, in the old astronomy. This step had a clear meaning in the Copernican system, for omission of it was equivalent to shifting the observer to the sun. That meaning gave Galileo immediately the key to predictions of satellite eclipses when, in 1612, he first realised that such eclipses took place. Now, from a purely mathematical standpoint, the Tychonic system is identical with the Copernican. Yet in 1614, when Simon Mayr claimed priority of discovery of Jupiter's satellites and published tables of their motions a bit more accurate than those used by Galileo in 1612, Mayr admitted that he had never seen a satellite eclipse and offered no way to predict them. The same correction mentioned above was used in his tables, but to this Tychonian astronomer it had no meaning beyond that of an empirical adjustment of the 'sun's motion'. Mayr simply could not imagine himself as moving around the sun, or think in terms of what would be seen from the sun. Astronomers who would not regard the earth's motions as real were under a great handicap in understanding the motions they observed, regardless of 'mathematical equivalence'.

Outside his appendix to the *Sunspot Letters* Galileo published nothing about satellite eclipses, for two reasons. The first was that he hoped to sell a certain scheme of his for determining longitudes, so he kept his method of calculation secret. The second reason was that Galileo was prevented from ever again treating the earth's motions as real, by reason of the events about to be recounted, and so highly technical a matter as satellite eclipses could not be explained to lay readers of his *Dialogue* – let alone their explanation assuming the earth to be fixed.

Because we are now approaching the series of events which led directly to intervention of the church in a purely scientific issue, some preliminary comments may be useful. It is now often said that incontrovertible evidence for the earth's annual motion was not found until early in the nineteenth century, when the high precision of astronomical instruments first permitted detection of parallax in certain fixed stars. Direct evidence of the earth's daily rotation is similarly said to have awaited the Foucault pendulum in the mid-nineteenth century. Such statements are titillating, but they misrepresent the grounds of scientific conviction. No scientist in the nineteenth century had lingering doubts which he gave up at the time of those events. The issue of the earth's motions had been effectively settled for scientists by Newton's law of universal gravitation, which linked innumerable astronomical measurements and the occurrence of tides to the existence of the earth's two motions.

Someone might reply that granting this, it is still true that Galileo had no incontrovertible evidence for the Copernican system, since he died before Newton was born. Quite so, and indeed Galileo refrained from asserting that he had incontrovertible evidence. What he did have was a preponderance of evidence that linked together such things as the phases of Venus, satellite eclipses, planetary speeds and distances from the sun, and the existence of tides; that made these compatible with his terrestrial physics, and that showed Aristotelian cosmology and physics to be mistaken on various matters. There were still a great many remaining puzzles, but in that respect his situation was like Newton's, or for that matter like ours today. Nothing in science is immune from further discoveries. Science proceeds on preponderance of evidence, not on finality.

The preponderance of evidence known to Galileo indicated that the earth's motions were actual, and Galileo's belief in them was scientific, even though some of the evidence he relied on was later found to be scientifically inadequate.

With this background we can see why Galileo felt compelled to do all he could to prevent a mistake by the Church that would eventually tend to discredit its wisdom. The difficulty in this was that even persons expert in astronomy did not yet understand the weight of the evidence known to Galileo. It was impossible to explain that to theologians expert not in astronomy and physics but only in their erroneous Aristotelian counterparts, so any attempt to do so would be a waste of time. On the other hand, the founding fathers of Christianity had wisely separated faith from science, precisely to avoid crises of the same kind at the elementary levels that had already existed in their day. Galileo accordingly appealed to their authority in his zeal to save his Church from the very mistake that it actually made in 1616.

The gulf between incontrovertible evidence and preponderance of evidence separates Aristotle's science from Galileo's. It is still not well understood by Galileo's critics, who attribute to him grounds for confidence in his science that were not his, though he had no less confidence in his science than he had in his religion. The kind of evidence that Cardinal Bellarmine would have regarded as incontrovertible was not in the possession of Galileo, whose modern critics like to dwell on that fact. The preponderance of evidence in his possession, however, supported Galileo; and since it is that which always counts in science, his actions are best understood by keeping this in mind.

It should also be kept in mind that throughout the arguments of 1613–16, Galileo's purpose was not to prove one side of a scientific question, but to separate purely scientific questions from matters of faith in order that rational discussion might remain free. Many writers say that he wanted the Church to adopt the Copernican system, which is not only false but misses the whole point of the actual debate. Galileo did not want the Church to adopt either side of any scientific question and suppress the other as a matter of faith; if the Church were to suppress anything, he wrote, it should forbid any introduction of scriptural authority into debates that could be settled without it, by experience and reason alone. Such was the separation between religion and science desired by Galileo, who never questioned the right of the Church to intervene, but who strongly urged against its doing so. It was also the separation urged long before by St. Augustine, who pointed out that a heretic might be better informed than a Christian in astronomical matters; that Christians should not spend time studying astronomy that could be better spent in pious devotions; and that to stake Christianity on such matters would be improper.

Castelli was appointed in 1613 to Galileo's former chair of mathematics at the University of Pisa on Galileo's recommendation. The overseer of the University warned him, when he took the post, that he was not to teach Copernicanism. Castelli replied that Galileo had not only already advised him against that, but had said that in nearly twenty years of university teaching, he himself had never done so. It should be remembered that during Galileo's university career he had remained ignorant of the evidence for Copernicus that he now possessed, which in the opinion of his modern critics was still insufficient to carry proper scientific conviction. In their view, I should think it would follow that Galileo advised Castelli correctly, yet they prefer to say that he had long hypocritically concealed his scientific convictions. In any view, Galileo's advice to Castelli shows that he was no Copernican zealot at the end of 1613, when he had all the scientific evidence he would ever have.

The professors of philosophy who had combated with Galileo on floating bodies formed a league at Florence, led by Colombe, whose members undertook to refute anything Galileo said. Most of them were professors at Pisa, and they were hostile to Castelli as a disciple of Galileo from the beginning of his teaching there. Near the end of 1613 Castelli was invited to a court breakfast at which Cosimo, his mother the Grand Duchess Christina, his wife the Archduchess, and other members of the Medici family drew him into conversation about Jupiter's satellites, which in their honour Galileo had named 'the Medicean stars'. A professor of philosophy whose speciality was Platonism took occasion to tell Christina that

Galileo, who was not present, was wrong to say that the earth moved, because that contradicted the Bible.

After the breakfast Christina detained Castelli to speak as a theologian on this point, and in particular to discuss the biblical miracle of Joshua in which the sun was said to have been stopped. Castelli answered all the questions that were raised, and maintained that purely scientific matters should be decided on their own merits, from which the literal or metaphorical status of scriptural passages could then be determined. He sent to Galileo at Florence an account of the incident, upon which Galileo proceeded to write out his long *Letter to Castelli* in which he approved everything he had said and added more. This was the first letter in which Galileo argued that freedom of inquiry should be allowed by theologians in all matters that could be decided by appeal to 'sensate experiences and necessary demonstrations' alone. That phrase restricted the scope of science to things unrelated to salvation of the soul. No contradiction could exist between Nature, as the executrix of God's will, and the Bible as the repository of God's word. The bible often spoke metaphorically and always for the easy understanding of ordinary people. Its words are subject to interpretation, which should be left to theologians, while Nature speaks inexorably for herself.

One or two previous incidents involving the Bible had occurred, but this was the first serious one. A book published against the *Starry Messenger* had invoked the Bible against Jupiter's satellites when Galileo was at Rome in 1611, where the Jesuits had told him their low opinion of that book. In 1612 a rumour had reached Galileo that Niccolò Lorini, an elderly Dominican much liked by the Medici, had said that 'this fellow Ipernicus' seemed to contradict the Bible. The league at Florence had suggested getting a priest to attack Galileo but was reprimanded by a churchman, perhaps the archbishop of Florence, at whose home they had met. Now a professor of philosophy had spoken against Galileo's views to his employers, and Galileo took action.

Years of experience showed him that the best strategy was to separate questions of fact from matters of opinion. Thus when the location of the new star was in question, Galileo determined it by ordinary techniques of measurement. Cremonini appealed to its substance, which, being perishable, could not be celestial. Galileo treated that as an opinion contradicted by fact. In the matter of floating bodies, observational facts contradicted opinions about causes. Concerning sunspots, facts of measurement destroyed the opinion that they were far from the sun's surface.

Observations and measurements sufficiently defined the realm of scientific facts as the highest court of appeal, so far as Galileo was concerned. Whether such facts were or were not recognised by Aristotelian philosophers was a matter of no interest to him. A fairly thoroughgoing revision of their principles would be necessary to accommodate the facts, but that could be managed if anyone wanted to go to the trouble of working it out. If not, then science would proceed independently of philosophical opinions.

The Bible was quite a different matter. No contradiction of Holy Scripture could be permitted in science, any more than in other things. Fortunately the apparent contradictions between astronomy and the Bible were few in number, since the Bible did not attempt to teach astronomy as did the philosophers. Biblical interpretation was a matter of opinion – of expert theological opinion, which should accordingly be governed by astronomical and physical facts. Science could not proceed independently of expert theological opinion, but agreement between them could easily be assured.

Such was Galileo's position, for which the most solid precedents existed. The early Church Fathers had recommended against any linkage of Christian faith with matters that were irrelevant to salvation, especially matters which required detailed study that would interfere with time better spent in devout meditations. The Council of Trent had fixed upon unanimous agreement of the Church Fathers as a basis of biblical interpretation, and none of them had advised making worldly knowledge depend on faith. Hence Galileo felt secure in his position.

For about a year there were no further developments adverse to Galileo, though Castelli was repeatedly harassed at the university. Then suddenly, in December 1614, a young Dominican named Thomas Caccini, a member of the same convent as Lorini, devoted a sermon from the pulpit of a principal church in Florence to a denunciation of mathematicians in general and of the Galileists in particular, his text being the miracle of Joshua. That same text had been the subject of a particularly long discussion in Galileo's *Letter to Castelli*, where it was treated *ad hominem* against the Aristotelians so as to show that for the effects described in the Bible to follow, the words 'Sun, stand thou still' could not be literally reconciled even with the cosmology accepted by all philosophers of the time.

Caccini's sermon at Florence created a stir elsewhere in Italy. It was not his first venture into sensationalism, for he had been already reprimanded for an earlier indiscretion from the pulpit at Bologna. Caccini had his eye on a desirable appointment at Rome and seems to have believed that his attack on the Galileists (for indeed some of Galileo's young disciples at Florence were so styling themselves) would help him to get it. His own brother took the exact opposite view and strongly urged him to desist from such tactics. A Dominican father at Rome wrote to Galileo apologising for the misbehaviour of Caccini as a member of his order. Prince Cesi suggested a concerted protest by professors of mathematics at all universities, though without bringing up the Copernican issue specifically, and thought that the archbishop of Florence should be asked to rebuke Caccini. Castelli, having experienced continual hostility from professors and administrators at the University of Pisa, was not surprised at Caccini's action in denouncing mathematics, the least controversial of all subjects; he wrote to Galileo that 'these attacks are not the first and they will not be the last'.

When Caccini's sermon was heard of at Pisa, the ruling family was in residence as was their custom at this time of year, and Lorini happened to be there too. He expressed regret that Caccini had gone so far, upon which Castelli showed him Galileo's letter of the previous year. Lorini copied most of this and took it with him to Florence; having discussed its contents with others of his convent, he forwarded it to the Roman Inquisition for investigation but without any accusations against Galileo or his followers. Galileo learned of this, or guessed that Lorini had made his copy for such a purpose. Fearing that his own words might have been altered, he recovered the original from Castelli and sent an exact copy to Piero Dini, a churchman he had met at Rome, asking him to show it to the Jesuits and, if possible, to Cardinal Bellarmine.

The partial copy had been read at a regular meeting of the cardinals of the Inquisition, who then asked the archbishop of Pisa to secure the original from Castelli and send it to Rome. In due course a qualified theologian wrote a report on it for official action, finding only an occasional word or phrase in it to be perhaps ill-advised and pronouncing it generally to be theologically unexceptionable. Caccini went to Rome to offer testimony against Galileo, and after he had been heard, two other witnesses he had named were sent for. The case was then closed for want of evidence that Galileo had said or written anything offensive to the Church. It is of interest that the Inquisition raised no objection even to this sentence in Galileo's *Letter to Castelli:*

> Scripture being therefore in many places not only liable to, but necessarily requiring, expositions different from the apparent meaning of the words, it seems to me that in physical disputes it should be reserved to the last place. (GW 225)

It is apparent that theologians were not seeking some pretext to censure Galileo, let alone to intervene in scientific issues. The problem was rather with intriguing personal foes of Galileo, and an ambitious priest, than with responsible Church officials. In mid-1615 Galileo expanded his *Letter to Castelli* into the much lengthier *Letter to Christina*, citing St. Augustine and other authorities whom he was certain that the Church would officially follow if, as he now expected, action were to be taken on the question of banning Copernican books. He corresponded with Dini and other friends at Rome, all of whom assured him that the stir raised by Caccini's sermon had died down and that nothing else seemed to be going on. Bellarmine in particular had said that there was no thought of banning Copernicus's book, but at most of removing some passages in it and leaving his astronomical hypothesis intact.

At this time a Neapolitan theologian, the Carmelite father P. A. Foscarini, published a book reconciling Copernican astronomy with the Bible, passage after passage, and came to Rome to debate the issue with anyone who wished. He submitted a copy of his book to Bellarmine and received a courteous reply in which Galileo as well as the author was mentioned by name. So long as they treated motion of the earth hypothetically, the cardinal said, they did well. But to assert the earth's motion as actually true would entail more, and more difficult, biblical reinterpretations than they supposed, and they were advised not to precipitate official action in that way.

Bellarmine's recommendation could have been accepted by Galileo without greatly impeding the progress of astronomy, and many scholars believe that Galileo should have accepted it, not only for his own safety but as sound science. His refusal to do so is ordinarily taken as evidence of excessive Copernican zeal. An unreasoning enthusiasm for a still unproved scientific system would indeed explain Galileo's next actions. But it is not necessary to assume such enthusiasm on his part to explain those actions, nor do they fit that assumption very well. For what Galileo did next was to set out at great length his arguments that the Catholic faith should not in any way depend on facts of science. There could be no contradiction between the Bible and science, and what should be done was to make that clear. The Bible should not even be construed as favouring, let alone adopting, one astronomy against another, or as requiring scriptural reinterpretation to accommodate anything that science might ever prove. As Galileo wrote in his old age, in his own copy of the later *Dialogue*:

> Take note, theologians, that in your desire to make matters of faith out of propositions relating to the fixity of sun and earth, you run the risk of eventually having to condemn as heretics those who would declare the earth to stand still and the sun to change position – eventually, I say, at such a time as it might be physically or logically proved that the earth moves and the sun stands still. (D iii)

Of course Galileo's *Letter to Christina* was not addressed directly to theologians, though it was intended for their eyes. It would have been improper for Galileo, as a layman, to address them in writing offering his advice on a matter of their expert judgement. The way to make sure that everything would be considered was to circulate in manuscript his personal views, and to go to Rome where he could clear them with friendly officials. Another reason for his going to Rome was that suspicion had been cast on him privately and publicly by Caccini, who had since been examined by the Inquisition, and Galileo wanted to clear his name of any charges. From what is known about his crucial Roman visit at the end of 1615, Galileo also argued the merits of the Copernican astronomy, in many gatherings, but that does not appear to have been a principal purpose of his going. Public opinion, or even informed opinion, about the truth of Copernicanism would have little weight in the deliberations of theologians which concerned Galileo.

The Tuscan ambassador at Rome warned the Grand Duke that Pope Paul V was so hostile to intellectuals of every kind that they had learned to conceal their true views; this, he said was 'no time to come to Rome and argue about the moon'. Nevertheless Cosimo authorised Galileo's journey and provided for his lodging at the Trinità del Monte, the Tuscan embassy.

The ambassador's characterisation of Pope Paul V was probably a reflection of actual general nervousness among intellectuals at Rome at the time, for which there were reasons. A principal area of contention between Catholics and Protestants was freedom to interpret the Bible, which meant that any new Catholic interpretation could be used by Protestants as leverage: if one reinterpretation could be made, why not wholesale reinterpretations? A dispute between the Dominicans and the Jesuits over certain issues of free will was still fresh in the pope's mind, as he had to take action in 1607 to stop members of the two great teaching orders from hurling charges of heresy at each other. These things suggest that Paul V, if not temperamentally anti-intellectual, had formed a habit of nipping in the bud any intellectual disputes that might grow into factionalism within the Church and become a source of strength for the contentions of the Protestants.

At Rome Galileo argued his astronomy against Aristotelian cosmology in various places and before various groups. His old Paduan friend, Antonio Querengo, reported in letters that although Galileo won few converts to his views, he utterly demolished the position of his opponents. Galileo found it hard, however, to meet personally with some officials to discuss theological issues, and was obliged to deal with them through intermediaries. Early in 1616 he wrote out at length his tide theory, based on the earth's motions, for Alessandro Cardinal

Orsini. But when Orsini approached the pope, he was told instead to persuade Galileo to desist from further argument lest the Inquisition (which was entirely at the pope's command) be set in motion against him.

Cardinal Bellarmine, consulted by the pope, advised that the propositions in dispute be submitted to the theological qualifiers who normally decided such issues. Galileo would then be notified of action based on their ruling. That was the procedure adopted; the two propositions submitted and the qualifiers' opinions on them were as follows:

1. That the sun is in the centre of the world, and totally immovable as to locomotion.

   Censure: All say that the said proposition is foolish and absurd in Philosophy, and formally heretical inasmuch as it contradicts the express opinion of Holy Scriptures in many places, according to the words themselves and according to the common expositions and meanings of the Church Fathers and doctors of theology.

2. That the earth is neither in the centre of the world nor immovable, but moves as a whole and in daily motion.

   Censure: All say this proposition receives the same censure in Philosophy, and with regard to Theological verity it is at least erroneous in the faith. (OP xix 321)

It is of interest that in both cases the censures had been made to hinge on the status of the propositions in Philosophy. Of no less interest is the phrase 'foolish and absurd', not *false*. Nothing was said of astronomy, it being simply assumed that astronomers were under the jurisdiction of philosophers. Had a panel of astronomers been asked for an opinion, it would doubtless have supported the qualifiers, as Galileo would have been outvoted on any panel of astronomers conceivable at the time. Yet if the qualifiers *had* consulted a panel of astronomers, and had so stated in rendering their censures, historians would to this day have blamed astronomers rather than theologians for the decisions taken.

It is a curious fact that historians have not blamed philosophers rather than theologians for the decision taken against freedom of scientific opinion in astronomy. Yet philosophers alone urged the intervention of theologians, confident that they would be on their side. Galileo appears from his letters at this time to have been equally confident that official action would take *neither* side; that theologians responsible for the future of the Church would decline to make an article of faith out of a disputed astronomical question. The shift of responsibility for interpretation of the Bible from theology to Philosophy took him by surprise. That ordinary people, and even some priests, would take the literal words of the Bible as supporting Aristotle's cosmology was to be expected. For the Church to adopt that view officially seemed to Galileo an unprecedented action. In his own copy of the later *Dialogue* he made these notes:

> On the matter of introducing novelties
> Does anyone doubt that from wanting minds created free by God to make themselves slaves of others' will, most serious scandals will be born?
> and wanting people to deny their own senses and subject them to the rule of another;
> and allowing persons entirely ignorant of a science as judges over those knowing it, so that by the authority conceded to them they are empowered to have things their way: These are novelties capable of ruining republics and subverting states. (OP vii 540)

This was written years later, but it represents Galileo's view of what had happened. The charge of innovation had been a favourite one against him by conservative professors, to whom novelty was anathema. By 'others' he referred to the Peripatetics, and by 'another', to Aristotle, for he used the same language of enslavement in the *Dialogue* itself. Concession of power to them by theologians seemed to him an innovation, and one capable of overthrowing good government. Since republics and states were not literally concerned, he doubtless had in mind the Church, though he could not bring himself to put so dreadful a prediction down on paper.

It is important to notice that the qualifiers did not simply justify their censures by saying 'It is foolish and absurd to say that the sun is fixed' or 'All say that the sun moves'. That would have been to take full responsibility on themselves. But the question put to them was not whether some way of talking was foolish; it was whether a certain way of speaking about astronomy was in conflict with the Bible. That necessarily raised the question whether relevant biblical passages spoke metaphorically or were intended to convey astronomical truths. The qualifiers decided the question by appealing to Philosophy, and that constituted a shift of responsibility for biblical interpretation in matters that could be settled by science. St. Augustine, or Aquinas, would have declared that the true sense of the Bible supported whichever astronomical hypothesis was verified in Nature, whether or not either had yet been verified beyond doubt by astronomers. That is what Galileo expected responsible theologians to say; instead, they said that the Bible supported the prevailing school of philosophers.

The recommendations of the qualifiers were read in the weekly meeting of cardinals of the Inquisition on 24 February 1616. The pope then asked Bellarmine to notify Galileo that he could no longer hold or defend the propositions censured. If Galileo resisted, the Commissary General of the Inquisition was to order him, in the presence of a notary and witnesses, that he must not hold, defend, *or teach* the propositions, lest the Inquisition proceed against him. The intention of the twofold order was clear: if Galileo submitted without protest, *no personal order* was to be given that was more stringent than the general instruction to all Catholics which would be published officially.

Everything later hinged on the word 'teach'. If Galileo was personally commanded not to teach the Copernican system in any way, he was bound not even to describe it. If no such command was given to him, he remained free to discuss the Copernican system, as could any Catholic, provided that he did not hold it to be true or defend it as more than a mere astronomical hypothesis. To weigh all the arguments on both sides of an issue was not regarded as defending one side against the other. It will be seen that Galileo was eventually ordered to stand trial only because of this word 'teach', and that his whole defence depended on showing that no personal command to him had been disobeyed.

What actually happened when Galileo met with the Commissary of the Inquisition has been debated for more than a century. The basic source of disagreement goes back to two documents bound into the later proceedings against Galileo. The first, drawn up by or for a notary but left unsigned, recounts a single meeting at the residence of Cardinal Bellarmine during which Galileo was told by the cardinal of the decision against motions of the earth and stability of the sun, which therefore could not be held or defended, following which he was then immediately forbidden by the Commissary in the name of the pope to hold, defend, or teach in any way, orally or in writing, the propositions named. The second document is an affidavit given to Galileo by Cardinal Bellarmine to the effect that Galileo was told no more than that the two propositions had been censured and that he must no longer hold or defend them.

Partisans of Galileo have branded the first document as a falsification of the events, while partisans of the Church have suggested that Bellarmine's affidavit was no more than a kindly equivocation designed to protect Galileo in his relations with the Tuscan ruler who employed him. Since neither tampering with Inquisition records nor prevarication by Cardinal Bellarmine is at all probable, both documents should be regarded as genuine, if that is possible. An otherwise unexplained statement by Galileo at his later trial suggests the sequence of events to have been substantially as follows.

On the morning of 26 February 1616 Cardinal Bellarmine sent two officers of arrest to summon Galileo to his residence. The Commissary of the Inquisition arrived uninvited with a notary and some Dominican fathers, to make sure that the liberal Jesuit was not lenient with any protest by Galileo. The cardinal resented this, but could hardly exclude them. Now, it was his custom to greet every visitor at the door, hat in hand, and it is known that when Galileo arrived the cardinal said something to him before they joined the others. Probably Bellarmine told Galileo that he was expected to offer no objection to anything that was said to him. A word to him would suffice, for Galileo was accustomed to dealing with cardinals and knew how things were done.

Rejoining the others, the cardinal took his chair and notified Galileo officially of the decision. But the Commissary had seen that something had already been said to Galileo and, suspecting that Bellarmine had told Galileo not to object, he took that as enough to release him from any previous restriction. Accordingly, without allowing Galileo time to reply, he delivered his personal injunction, in the name of the pope but even more strongly than the pope had authorised. Galileo then acquiesced. All this was recorded by the notary.

Bellarmine escorted Galileo to the door, urging him to pay another visit before returning to Florence. He then reprimanded the Commissary, privately, for an action contrary to the pope's express instruction. The notary's statement could not be signed by him in such circumstances, and he would instruct Galileo to treat everything except the cardinal's legal order as never having happened, no threat against him having been intended by the pope so long as he strictly obeyed that order.

At the next meeting of the cardinals of the Inquisition, Bellarmine reported that Galileo had been advised of the pope's decision and had acquiesced in it. The Commissary was also present at this meeting, where the minutes show that he added nothing. On 5 March a decree was issued which placed on the Index of Prohibited Books those works in which motion of the earth and stability of the sun were treated as real or were reconciled with the Bible. Foscarini's book was absolutely forbidden, while that of Copernicus and a certain commentary on the Book of Job were suspended pending correction. This meant the removal of passages relating to scriptural reconciliation or going beyond merely hypothetical treatment of the Copernican assumptions.

Bellarmine soon spoke again with Galileo and also informed the pope as to what had actually taken place. For within a few days Galileo was fully informed of the corrections to be made in *De revolutionibus*, not published until 1620, and was also granted an audience with the pope. There he was assured that the intrigues of his foes as well as his own upright conduct were known, and was told that as long as Paul V lived he had no cause for worry.

Letters from friends at Pisa and Venice showed that Galileo was rumoured to have been punished and forced to abjure. Galileo took these to Bellarmine in May, asking for some concrete evidence that would convince his employers of the untruth of such stories. The cardinal wrote his affidavit, and Galileo returned to Florence.

Having been effectively silenced on Copernicanism, Galileo turned his attention to other matters, and first to a practical one. At the time of his discovery of satellite eclipses in 1612, Galileo hit on a scheme to use them for more accurate determinations of longitude, for which ordinary eclipses had long been used. That in turn suggested to him a way in which ships at sea could determine their approximate longitudes by using Jupiter as a kind of celestial clock with its satellites as pointers. The project had been presented to the Spanish government through the Tuscan ambassador, but had languished. Galileo now began perfecting his tables of satellite motions, bringing them to remarkable accuracy in 1617. His longitude scheme was never adopted by Spain, but towards the end of his life the Dutch government offered him a handsome reward for it.

Next Galileo returned to his Paduan work on motion, intending to complete his long-neglected treatise on that subject. But just as he began this, in the autumn of 1618, three comets appeared and excited much attention. Galileo's opinion was sought, while numerous books about comets were published. One of these was printed as representing the opinions of the mathematicians at the Jesuit College in Rome, the anonymous author being Orazio Grassi.

Mario Guiducci, who had been assisting Galileo with his treatise on motion, had recently been elected head of the Florentine Academy and needed a topic for his inaugural address. It was decided that he should discuss comets along the lines taken by Galileo in conversations with friends. Part of Guiducci's lecture, subsequently published in book form, was critical of the Jesuit position on two matters. First, Grassi had adhered to Tycho's idea that a comet was a sort of quasi-planet, created and destroyed in one of the planetary orbits. Galileo, like Kepler, recognised that a comet's visible path was more nearly a straight line than a circle. The other Jesuit position had to do with the magnifying properties of the telescope, wrongly conceived. Guiducci's criticisms much offended the Jesuits, who correctly took them to be Galileo's. Writing with the pseudonym Lothario Sarsi, Grassi published a slashing attack directly against Galileo in which he went so far as to accuse him of surreptitiously upholding Copernicanism in accounting for apparent curvature of cometary paths.

The Linceans at Rome urged Galileo to reply. This he could safely do, since comets were not regarded as celestial by the

Aristotelians and Copernicus had said nothing about them. Published in 1623, *The Assayer* outlined Galileo's conception of scientific reasoning in contrast with the tiresome logical quibbles that satisfied natural philosophers. It contained some passages now celebrated but often cited out of context, as for example the closing lines of this statement:

> In Sarsi I seem to discern the firm belief that in philosophising one must support oneself on the opinion of some celebrated author, as if our minds ought to remain completely sterile and barren unless wedded to the reasoning of someone else. Possibly he thinks that philosophy is a book of fiction by some author, like the *Iliad* or *Orlando Furioso* – productions in which the least important thing is whether what is written in them is true. Well, Sarsi, that is not how things are. Philosophy is written in this grand book the universe, which stands continually open to our gaze. But the book cannot be understood unless one first learns to comprehend the language and to read the alphabet in which it is composed. It is written in the language of mathematics, and its characters are triangles, circles, and other geometric figures, without which it is humanly impossible to understand a single word of it; without these, one wanders about in a dark labyrinth. (D&O 237–8)

The final sentences are often said to mean that Galileo, like Plato, cared not about Nature but about a mathematically perfect world behind or above Nature. Yet Galileo spoke here of mathematics as a language necessary for understanding nature, not as an end in itself. Mathematical regularity stood in contrast with:

> the 'sympathy', 'antipathy', 'occult properties', 'influences', and other terms employed by philosophers as a cloak for the correct reply, which would be 'I do not know.' That reply is as much more tolerable than the others as candid honesty is more beautiful than deceitful duplicity. (D&O 241)

In another passage Galileo distinguished sensations from properties of external physical bodies:

> I think that tastes, odours, colours, and so on are no more than mere names so far as the objects in which we locate them are concerned, and that they reside only in consciousness. If living creatures were removed, all these qualities would be wiped out and annihilated. (D&O 274)

Some say that by removing, say, the colour red from objects we describe as red, Galileo divorced humanity from science, though it would be equally accurate to say that he directed special attention to sensation and consciousness. His aim was to dispel the idea that words have the kind of power with which they are endowed by philosophers:

> If their opinions and their voices have the power to call into existence the things they name, then I beg them to do me the favour of naming a lot of old hardware I have about my house 'gold'. (D&O 253)

The distinction of sensations from external physical phenomena was later made an important part of the empiricist philosophy by John Locke, and it is usually referred to as the separation of primary from secondary qualities. Neither that terminology nor empiricist philosophical views belonged to Galileo, though both are often ascribed to him on the basis of remarks found in *The Assayer* The general idea was of much earlier origin, being found in Lucretius and having probably been associated with Greek atomism from its beginnings. Galileo was neither an empiricist nor a rationalist in the senses those terms came to have in philosophical disputes after Descartes and Locke. His science required simultaneously sensible experience and necessary demonstration; Galileo did not accord greater 'reality' to one than to the other, nor did he regard sensation as nonexistent or less important than external physical phenomena. He did wish to distinguish things that were different in kind but were frequently confused by philosophers.

The Lincean Academy was about to publish *The Assayer* at Rome just when Maffeo Barberini became Pope Urban VIII. Because he was a Florentine, an intellectual, and an admirer of Galileo, the Linceans decided to dedicate the new book to him. Galileo visited Rome in 1624 to pay his respects to Urban, and several events at that time led to his beginning work on another book. A German cardinal told the pope that the edict of 1616 had lost the Church some prospective converts, and later this pope said that if it had been up to him, that edict would not have been issued. Galileo outlined his tide theory to Urban, which he had wished to publish but which depended on assuming the Copernican motions of the earth.

Italian leadership in science would be lost if the edict were too strictly construed. In the six audiences he was granted with the pope during this visit, Galileo appears to have obtained his permission to publish his tide theory, making it clear that the earth's motions were taken only hypothetically and could not be proved real by earthly experiments or celestial observations. In that way the Church, Italian primacy in science, and Galileo's own interests could all be served without any need to rescind the edict, which Urban would not do.

This appears to have been the understanding when Galileo left Rome with many tokens of the pope's esteem and affection. But he had not mentioned the incident in 1616 which Cardinal Bellarmine had told him to treat as if it never had happened.

5 The *Dialogue* and the Inquisition

From 1624 to 1630 Galileo was intermittently at work on his book, which at the last moment he was instructed not to call 'Dialogue on the Tides' because that would stress a physical argument for motions of the earth. It was a reasonable instruction in view of the traditional astronomical meaning of treating planetary motions hypothetically only, and leaving all physical considerations out of account, so Galileo changed the title to *Dialogue Concerning the Two Chief Systems of the World – Ptolemaic and Copernican.*

The dialogue form had been chosen for various reasons, among which was the fact that during the sixteenth century that form had become very popular for books designed to educate the public. The master–pupil conversations that first appeared for such purposes tended to be dull catechisms, so Galileo's dialogue introduced in effect two experts who vied for the support of a third, uncommitted, participant. Another reason for writing in dialogue form was that the author could detach himself from commitment to views that might be objectionable. One spokesman principally represented Galileo, who himself appeared in the book only as 'our friend', or 'the Academician', or the like, when he wished to assert his personal claim to or responsibility for certain things.

Galileo made his principal spokesman Filippo Salviati, who had died suddenly on a visit to Spain in 1614. The expert Aristotelian was called Simplicio, after a distinguished ancient Greek commentator on Aristotle. His arguments were patterned on those of Cesare Cremonini and Ludovico delle Colombe. The interested layman was represented by Giovanfrancesco Sagredo, who had died in 1620, and we have the word of a Venetian friend of his that Galileo accurately recaptured his mind and style.

The dialogue was divided into conversations during four 'days' which the interlocutors had set aside for exploring the relative merits of the old and new astronomies. On the first day, the Aristotelian division between celestial and elemental substances and their associated motions opened the discussion. This fundamental tenet of natural philosophy came in for criticism partly on logical grounds and partly with reference to new astronomical knowledge since Aristotle's time. From the logical standpoint Aristotle was accused of having often assumed what was to be proved, and of having made unrecognised and unjustified assumptions. Chief among the new discoveries discussed were features of the moon's surface and the continually changing illumination of mountains and craters.

The second day was devoted mainly to showing that no standard argument against the earth's daily rotation is conclusive. Relativity of motion and conservation of motion constituted Galileo's main weapons, the arguments being mainly physical rather than astronomical.

The third day concerned the earth's annual motion around the sun, and of course certain phenomena involving both the daily rotation and the annual revolution. Among the latter was the cyclical change of sunspot paths throughout the year. That was easily explained by assuming both Copernican motions, but was very complicated, and dynamically incredible, if all necessary motions were located in the sun. Galileo's inclusion of this argument infuriated Christopher Scheiner, as will be seen,

In presenting the Copernican system to readers of the *Dialogue*, Galileo not only ignored the elliptical planetary orbits that had been introduced by Kepler, but oversimplified the astronomy even of Copernicus by treating the sun as if it were at the centre of all planetary orbits. This has led to much modern criticism by scholars who ignore the purpose of the *Dialogue* and speak as if Galileo intended it to be a textbook of astronomy. His purpose was to break down resistance to motions of the earth, in order to use those to explain the tides. He was in fact prevented by the 1616 edict from dealing with those motions except hypothetically; all he could do was to show the invalidity of all arguments that had been offered to prove the earth to be at rest. In the third day, dealing with the annual revolution of the earth around the sun, he argued that this offered a simpler scheme for astronomers, and it sufficed to illustrate this by the first diagram in Copernicus's book without going into the further technical details.

As to Kepler's elliptical orbits, which were indeed the true beginning of modern astronomy, a good deal of misunderstanding prevails. The ellipticity is very slight, though it is the key to the mathematical laws of planetary motions. As a first approximation, circular orbits around the sun serve very well, and we have already considered Galileo's attitude towards approximation in science. Hence there would be no real mystery about his omission of Kepler's ellipses from the *Dialogue*, even if Kepler had not been a German Protestant whose works were on the Index of Prohibited Books and if his astronomy had not been virtually impossible to explain to lay readers of Galileo's time.

In the fourth day Galileo dealt with the tides. Short of invoking miracles, he began, there is no way to explain great and recurring motions of large seas on an absolutely fixed earth. That is correct, and it follows that any scientific explanation of the tides must involve motion of the earth. Hence, though Galileo's tide theory was entirely inadequate, he had put his finger on the one well-known ordinary physical phenomenon that indeed requires the new astronomy. That was no mere happy accident, for Galileo reasoned from a situation that would disturb large seas, according to later physics, though not enough to account for anything like the tides we actually observe.

Galileo's theory of the tides is so thoroughly misrepresented in most books as to appear simply absurd. Galileo offered two basic causes, one for continuous disturbance of the seas and the other for the period of tides in the Mediterranean, which could not be deduced from the period of continuous disturbance. It is now customary to ignore one of Galileo's two causes, deduce from the other what he said could not be deduced, and then to assert that he made a silly mistake by reason of his Copernican zeal. Galileo's tide theory was incorrect but scientific, as were the very different tide theories of Newton and Laplace; a reasonably correct theory was developed only late in the last century.

There were difficulties in obtaining a licence to publish the *Dialogue*, and soon after it was licensed at Rome the sudden death of Prince Cesi disorganised the Lincean Academy which had intended to publish it. Eventually a second licence was given to publish the book at Florence, where it appeared in

March 1632. An outbreak of plague delayed the sending of copies to Rome.

Suddenly in August an order came from the Roman Inquisition to stop all sales, and Galileo was summoned to stand trial. Cosimo was dead, but the young Grand Duke Ferdinand protested strongly against such treatment of the author of a duly licensed book. It was no use; Urban VIII was adamant and very angry. Even Galileo's serious illness delayed things only for a time, though in the opinion of doctors who examined him for the Florentine Inquisition he could be moved only at peril to his life. It was winter by then, with delays along the road for quarantine because of the plague, and Galileo was nearing seventy; nevertheless he was told to come to Rome or be brought in chains and pay the expenses of arresting officers sent from there.

What had happened was that Urban VIII had been shown the unsigned notary's memorandum of 1616. The pope had no reason to disbelieve it, and since Galileo had never told him of any personal injunction not even to discuss Copernicus, it appeared to Urban that a legal order had been disobeyed. Who had dug out this document, which should have been destroyed as of no legal value, is not known. Informed opinion at Rome was that Scheiner had been responsible, which is highly probable. He had published in 1630 a huge book on sunspots which contained, among other things, a violent attack on Galileo and details about the annual change in sunspot paths. Scheiner assumed that Galileo's argument in the third day of the *Dialogue* (which in fact had been licensed before Galileo saw Scheiner's book), was based on information obtained from his book and applied to support Copernicus. Having moved to Rome in 1624, Scheiner was in a position to stir up the Inquisition. In any event the document found in its records convinced the pope that Galileo had deliberately deceived him.

Galileo arrived at Rome in February 1633 and was lodged with a new and very friendly Tuscan ambassador who was able to tell Galileo enough for him to know that all that was really in question was the 1616 meeting at Bellarmine's residence. The ambassador, who knew at first hand the pope's rage, was surprised at Galileo's confidence as to the outcome. Neither he nor anyone else living (except Galileo) knew of Bellarmine's affidavit.

It was not until 12 April that the trial began. After a series of questions about the writing, licensing, and printing of the *Dialogue*, the qualifiers' ruling of 1616 was brought up; Galileo was asked who had told him about it, and he replied:

In the month of February 1616, Cardinal Bellarmine told me that since the opinion of Copernicus absolutely taken contradicted Holy Scripture, it could not be held or defended, but that it might be taken hypothetically and made use of. In conformity with this I have an affidavit of the same Cardinal Bellarmine made in the month of May, on the 26th, 1616 ... of which affidavit I present a copy. ... The original of this affidavit I have with me in Rome, and it is entirely written in the hand of Cardinal Bellarmine. (GW 346)

The prosecutor entered this exhibit. He then asked whether others had been present, and whether anyone else had given Galileo a precept of any kind. Galileo said that some Dominican fathers, whom he did not know, had been there, and went on:

As I recall, the affair came about in this manner: One morning Cardinal Bellarmine sent for me, and he told me a certain particular which I should like to speak to the ear of His Holiness before that of anyone else; but in the end he told me that the opinion of Copernicus could not be held or defended, as contrary to the Holy Scriptures. As to those Dominican fathers, I do not remember whether they were there first, or came afterwards; nor do I recall whether they were present when the Cardinal told me that the said opinion could not be held. And it may be that some [personal] precept was made to me that I might not hold or defend the said opinion, but I have no memory of it, because this was many years ago. (GW 346–7)

The prosecutor then read to Galileo the precept, which included the words 'nor teach in any way'. Galileo stood his ground, saying that he recalled no more than Bellarmine's admonition and that he had always relied on the affidavit – which said that he 'was only told of the declaration made by His Holiness and published by the Congregation of the Index' that motion of the earth 'is contrary to Holy Scripture and therefore may not be defended or held'.

In due course Galileo was asked to produce the signed original in Bellarmine's hand, which he did. No signed document was ever found to support the memorandum on which the Inquisition had based its charge, so on the only substantive issue raised Galileo had won by the rule of best evidence. No scientific question was raised at the trial; the charge was 'vehement suspicion of heresy', for which it was sufficient to have disobeyed an official order, whether or not any heresy had been uttered.

Galileo could not be acquitted without damage to the reputation and authority of the Roman Inquisition, so it was privately arranged that he should admit to some wrongdoing and submit his defence, with the understanding that he would be treated leniently. Galileo acknowledged in writing that he had re-read his *Dialogue* and had found places in it where he had gone too far; he then pleaded any man's natural vanity with respect to his own arguments and denied any sinister intent. Still expecting a light sentence, he was crushed by his condemnation to indefinite imprisonment.

During his stay in Rome Galileo had been invited to visit Archbishop Ascanio Piccolomini of Siena after the trial. The ambassador contrived to have Galileo's sentence commuted to custody of the archbishop, whose humanity and understanding literally saved Galileo's life and sanity. He managed to get Galileo's mind back on science, encouraging him to begin writing his long-planned treatise on motion. The archbishop already knew of this through his teacher of mathematics, Buonaventura Cavalieri, a pupil of Castelli and friend of Galileo.

Galileo's eldest daughter, Virginia, had entered a Franciscan convent at Arcetri in 1616, taking the name of Sister Maria Celeste. Galileo was very devoted to her, though his visits to the convent had been less frequent than either of them wished. His home was then in Bellosguardo, a considerable distance from Arcetri. Frequent illnesses made it difficult for him to make the journey, and also greatly distressed his daughter because of her inability to look after him. In 1631 he had taken a villa in Arcetri adjoining the property of the convent, to which he was now very anxious to return. His daughter's letters show her to have been a woman of unusual intelligence and sensitivity who never wavered in her loyalty either to her father or to her religion, even during the trial. Though her situation at the convent must have been difficult at that time, it appears that the other sisters shared her joy at Galileo's release from imprisonment at Rome. Her letters redoubled his desire to return, as when she wrote:

> There are two pigeons in the dovecote waiting for you to come and eat them; there are beans in the garden waiting for you to pick them. Your tower laments your long absence. When you were in Rome, I said to myself: 'If he were only at Siena!' Now that you are at Siena I say: 'If only he were at Arcetri!' But God's will be done. (PLG 266)

At the close of 1633 Galileo finally obtained permission to return to his villa, where he dwelt for the rest of his life under the surveillance of officers of the Inquisition.

Soon after his arrival Galileo suffered a serious hernia and requested permission from Rome to see doctors in Florence. This was refused, as reported to Galileo on the very day he last saw Sister Maria Celeste, gravely ill at the convent. Her death on 2 April 1634 dealt Galileo another blow from which he was long in recovering. For a time he did not even want to, writing to a friend later that month:

> I held off writing you about the state of my health, which is indeed sad. The hernia has returned larger than before; my heartbeat is cut into with palpitations; immense sorrow and melancholy [accompany] loss of appetite; hateful to myself, I continually hear calls from my beloved daughter; . . . in addition to which I am not a little frightened by constant wakefulness. . . . I have at present no heart for writing, being quite beside myself so that I neglect even replying to the personal letters of friends. (OP XVI 85)

6 The final years

While Galileo was first at Siena he had written to Sister Maria Celeste saying that his name had been removed from the book of the living, so deeply did he feel his condemnation by the Church. His letters to his daughter do not survive, but this is clear from her reply:

> Do not say that your name is struck *de libro viventium*, for that is not so, either in the rest of the world or in your own country. Rather, it seems to me that if your name and reputation were briefly under a cloud, they are now restored to greater fame – which is astonishing, since I know that no one is accounted a prophet in his own land. (PLG 265)

His daughter's cheering words related only to the joy and relief expressed at her convent over Galileo's escape from confinement in prison at Rome. They turned out to be prophetic, however, for in 1634 Galileo's *Mechanics* was translated into French by the Minim friar Marin Mersenne, long before it was printed in its original Italian. The next year his ill-starred *Dialogue* was printed in a Latin translation by Matthias Bernegger at Strasbourg, reaching in that form a far more cosmopolitan audience than the suppressed Italian text. And in 1636 the *Letter to Christina*, which had circulated only in manuscript copies, was printed together with a Latin translation, apprising all Europe of Galileo's position on the introduction of scriptural passages into purely physical questions. The Latin title of the book specified such questions to be those 'that can be evinced by sensate experiences and necessary demonstrations', so that Galileo's precise definition of the kind of science that should be exempted from theological censorship was clearly set forth for consideration by any scholar in Europe. The translator was Elia Diodati, long a correspondent of Galileo at Paris, and the printers were the Elzevirs, based in Holland.

Several books were published attacking the *Dialogue*, to which Galileo could not publicly reply. One such book, by a stubborn Peripatetic at Venice named Antonio Rocco, started Galileo on penning marginal notes in his copy and then writing out at length many replies and comments on additional sheets. He sent copies of these to the Servite friar, Fulgenzio Micanzio, who had been Sarpi's assistant and had succeeded him as theological adviser to the Venetian government on Sarpi's death in 1623. Micanzio had been a friend and admirer of Galileo when he was at Padua, and from this time became his constant correspondent and rendered him many valuable services. Micanzio wrote:

> I took to the villa your *Dialogue* and Rocco's book; nothing else. I have read both with pleasure, my mind reaching that state in which the eye is while watching the clown imitate the acrobat. . . . I can no longer abide speculative physics; it seems to me that, re-examining the Peripatetic principles as you have done, for me they all go up in smoke. (GW 362)

Among the many topics concerning which Galileo annotated Rocco's book was one, of great importance to mathematics, on which he had composed a treatise (now lost) before moving to Florence in 1610. The subject was the continuum, later dealt with by invention of the infinitesimal calculus. Galileo's geometrical analysis of the continuum had enabled him finally to understand uniform acceleration in fall, in 1608. He now applied a similar analysis to the structure of matter, as seen in the opening dialogue of his last and greatest scientific book, which he had begun to write at Siena. Called *Two New Sciences*, it was published at Leyden in 1638 by a member of the Elzevir family who had recently established an independent firm of his own. The writing of this book occupied Galileo from 1634 to 1637. It consisted of two more or less independent treatments of basic areas of physics, the structure of matter and the laws of motion, each written as two 'days' of conversation between the same interlocutors as in the *Dialogue*. With considerable skill, Galileo tied the two topics together, first laying a basis both in mathematics and in physics for his analysis of motion, which came at the end, by discussions in his opening section on the structure and resistance of materials.

The first half of this book was completed by mid-1635, and at the suggestion of a Medici prince Galileo gave a copy of the manuscript to a Florentine engineer who was leaving for Germany to serve the Holy Roman Emperor. The purpose was to find, if possible, a publisher for it who had not heard of the ruling at Rome that no book written or edited by Galileo, past or future, could be printed or reprinted. Galileo himself had not fully realised this until Micanzio wrote telling him what had happened when he had broached to the Venetian inquisitor the question of a licence for Galileo's new book, which had nothing whatever to do with any theological matter. He was told of the order and remonstrated that it could not mean, for example, that if Galileo wished to publish an edition of the Lord's Prayer he would be forbidden to do so. The inquisitor gravely showed him the wording, which indeed prohibited the printing of anything whatever that Galileo should ever write, or even just edit.

The engineer tried patiently for many months, in Germany and Poland, to find a publisher, but everywhere the Jesuits were alert to keep Galileo silent. Ironically, the only willing publisher finally found was a cardinal who had installed a press at his own home, but he died before work was started. Meanwhile Galileo had been advised by Diodati to try the Elzevirs, who had published the *Letter to Christina* with his Latin translation, and also the *Dialogue* in Latin. He also had had a request from a French mathematician to send him the manuscript, confident that he could find a French publisher. In the end Louis Elzevir, who was starting his own operation, visited Galileo at Arcetri and agreed to do the work, a part of which he received through Micanzio at Venice before leaving Italy, and the rest of which was sent via Micanzio to the printers at Leyden, in instalments. The dramatic story that Galileo had to smuggle out his manuscript past the vigilance of resident representatives of the Inquisition who kept him under surveillance is a fiction. In fact, from what little is known, it appears that those representatives became quite fond of Galileo while living with him and merely 'worked to rule', which consisted mainly of screening and reporting visitors to Galileo at Arcetri.

The *Dialogue* had been of considerable interest from the standpoint of physics, though it was probably read mainly because of its discussion of astronomical systems and its celebrity as a suppressed book. An English translation had been made, as Thomas Hobbes assured Galileo during a visit in 1634, but it was not published, probably because the Latin version of Bernegger made the text available to any interested scholar in England as elsewhere. Sir Isaac Newton read a later, independent, English translation published in 1661 and made some notes on it when he first considered the possibility of universal gravitation, in 1666. Galileo's law of falling bodies had been included in the *Dialogue*, though merely in passing and with a promise of development in some later book. Other things of importance to physics included Galileo's principle of relativity of motion, admirably developed later by Christian Huygens, and his concept of conservation of motion, extended by Descartes and established by Newton as the law of inertia in a way that Galileo had not believed scientifically legitimate. But all these things, sufficient in themselves, did not present anything like the main body of Galileo's physics, or approach the subject systematically as was done in *Two New Sciences*.

The first of those sciences was truly new; no one had previously discussed the structure of matter mathematically or had developed a theory of the breaking strength of materials. Engineers and architects had of course accumulated a vast practical knowledge of the latter. This was *techne* in Aristotle's sense, and the difference between that and useful science is illustrated by Galileo's work. Starting from the law of the lever and the assumption of uniform distribution of cohesion of parts of a solid, Galileo presented a series of theorems organising what was known and carrying it much further by mathematical deduction. Not the least interesting of his discoveries was that there is a limit to the size of anything made of the same materials and maintaining the same proportions:

> Nor could Nature make trees of immeasurable size, because their branches would eventually fall of their own weight; and likewise it would be impossible to fashion skeletons for men, horses, or other animals which could exist and carry out their functions proportionably when such animals were increased to immense height. . . . It follows that when bodies are diminished, their strengths do not proportionately diminish; rather, in very small bodies the strength grows in greater ratio, and I believe that a little dog might carry on

his back two or three dogs the same size, whereas I doubt if a horse could carry even one horse his size. . . . (TNS 127)

What happens in aquatic animals is the opposite of the case with land animals; in the latter, it is the task of the skeleton to sustain its own weight and that of the flesh, while in the former the flesh supports its own weight and that of the bones. And here the marvel ceases that there can be very vast animals in the water but not on the earth – that is to say, in the air. (TNS 129)

This provides an example of useful science as contrasted with practical knowledge. Mathematical organisation of existing knowledge so that further knowledge could be deduced from it was what principally interested Galileo, while others sought principles from which the whole of truth could be deduced by logic. Galileo's pursuits required work of a kind that left him little time, even if he had had the inclination, to indulge in philosophical speculation, with which he had become thoroughly impatient by the time he composed his last book. The calculations that had revealed to him the direction in which science was going to advance were very time-consuming; they fill hundreds of folios among his surviving working papers alone. Only such labours had put in his hands a preponderance of evidence against the established beliefs of natural philosophers. To grasp that evidence fully, others would have to do the same kind of work. Most people preferred grandiose promises, as Galileo had noted; but all he ever offered was modest and useful science.

The second science presented by Galileo was new in quite a different sense, as he stressed at its introduction. This was the science of natural motions, on which (as he said) a great many books had been written, but without attention to any of the basic properties of motion he investigated. Aristotle had made motion and change the basis of all physics, but no one had presented the law by which bodies are accelerated in natural descent, or had acknowledged the independent composition of motions that permitted accurate description of the paths of projectiles. The first of these underlay Galileo's third dialogue, and the other, his fourth and last. He had composed a fifth, on the force of percussion, but decided to withhold it from publication, being not entirely satisfied with it himself.

Galileo's science of natural motions was also useful science in the same sense as was his science of strength of materials. A great many theorems of only theoretical interest were included, but the basis was laid for mathematical treatment of many practical problems in physics. Galileo took care to have the Aristotelian, Simplicio, object that the horizontal plane is not truly a plane and that air resistance impedes motion, to which he had his own spokesman agree and name still other factors impossible to remove.

I admit that the conclusions demonstrated in the abstract are altered in the concrete. . . . [But] if such minutiae had to be taken into account in practical operations, we should have to commence by reprehending architects, who imagine that with plumb-lines they erect the highest towers in parallel lines [though these converge at the centre of the earth.] (TNS 223–4)

The mixture of *techne* and *episteme* that constituted useful science was rejected not only by philosophers of Galileo's day, but by others down to the present. Their continued separation since this time has been, however, a convenient fiction for purposes of analysis, and not a historical reality as it had been from the time of Aristotle to that of Galileo.

By the time *Two New Sciences* was printed, Galileo had become totally blind. During part of 1638 he was permitted by Rome, after much negotiation and many guarantees from the Florentine chief inquisitor, to live with his son in Florence for consultation with doctors, but forbidden to speak to others. Even to attend church services in Holy Week he had to get special permission to leave the house and was obliged to promise to converse with no one. Loss of sight was a particular affliction to Galileo, not only because he could not read or write any longer but because throughout his life a special talent for observation had led him to discoveries in physics as well as astronomy.

Alas, your friend and servant Galileo has for the last month been irremediably blind, so that this heaven, this earth, this universe which I, by my remarkable discoveries and clear demonstrations had enlarged a hundred times beyond what has been believed by wise men of past ages, for me is from this time forth shrunk into so small a space as to be filled by my own sensations. (PLG 283)

Towards the end of 1638 a young pupil, Vincenzio Viviani,

came to live and study with Galileo, serving him also as amanuensis. Viviani wrote the first fairly long biography of his teacher several years later, and though it contains demonstrable errors it is of special value because of personal anecdotes he had heard from Galileo during these last years of his life.

In 1640 Galileo dictated a long letter at the request of Prince Leopold de' Medici replying to parts of a book published by Fortunio Liceti, professor of philosophy at the University of Padua. Galileo had long been on friendly terms with Liceti, a prolific author of typical Peripatetic books explaining in orthodox Aristotelian terms comets, new stars, phosphorescent stones, and everything else for which scientists offered other explanations. In discussing phosphorescence he had misquoted and attacked Galileo's account in the *Dialogue* of the faint light seen on the nearly new moon as reflection of sunlight from the earth. Liceti heard of the letter and asked for a copy so that he might publish his refutations of it. Galileo agreed, though he wished first to revise and polish it because it had not been intended for publication. It is of special interest as reflecting near the end of Galileo's life a conciliatory attitude towards Aristotle, though by no means towards his self-styled disciples such as Liceti.

During their correspondence Galileo had said that he considered himself a better Aristotelian than those who complained that he attacked Aristotle. Very possibly Galileo took that line because in his heart he felt, though he could never say, that he had been a better theologian than those who had ruled against Copernicanism. At any rate, Liceti replied with some irony that it was most welcome news to him that Galileo claimed not to contradict Aristotle's teachings: 'I seem to have gathered the contrary from your writings, but it may be that on this matter I was mistaken along with many others of the same opinion.' Galileo replied:

> To be truly Peripatetic – that is, an Aristotelian philosopher – consists principally in philosophising in conformity with Aristotelian teachings . . . among which one is the avoidance of fallacies in reasoning. . . . As to that one, I believe I have learned sureness of demonstration from the innumerable advances made by pure mathematicians, never fallacious. Thus far, then, I am Peripatetic.
>
> Among the safe ways to pursue truth is the putting of experience

ahead of any reasoning, we being sure that any fallacy must reside in the latter at least covertly, for it is not possible that sensible experience is contrary to truth. This also is a precept highly esteemed by Aristotle which he placed far in front of the value of all the authority in the world. . . .

Those who clumsily adopt the above precept . . . would have it that it is good philosophising to accept and maintain any dictum or proposition written by Aristotle. To support these, they induce themselves to deny sense experience and give strange interpretations to Aristotle's texts. . . . I am certain that if Aristotle should return to earth, he would rather accept me among his followers, in view of my few but conclusive contradictions, than a great many other people who, in order to sustain his every saying as true, go pilfering conceptions from his texts that never entered his head. (GW 408–9)

Liceti had not only attacked Galileo's positions but had misquoted them in his book, attributing to him things he had never said, some of which had previously been ascribed to him by other Peripatetics. Galileo patiently corrected these, explained again his reasoning, and exposed innumerable faults in Liceti's arguments. But there was no hope whatever of winning academic philosophers over to sober and useful science; Liceti learned nothing about the dangers of blindly accepting authority and interpreting it to fit preconceived ideas. He refuted Galileo's letter in 183 sections of a tiresome book. Centuries later a few philosophers (the first being David Hume) came to look at science much as Galileo had, and a few others began to support ideas carelessly supposed to have been Galileo's, but most preferred to go on constructing what he called 'worlds on paper' remote from the sensible world.

It is in fact quite difficult to present the world of daily experience in a book printed on paper; Galileo was one of the few writers who succeeded quite well. This he did by a wealth of homely examples of things that have been noticed by everyone without their having thought about them. It is almost impossible to read one of Galileo's arguments in the *Dialogue* concerning the earth's reflected light which Liceti had denied, without calling to mind the real moon, real clouds and real mountains – not just the idea of 'moon' in astronomy and verbal abstractions about clouds and mountains:

SALVIATI: Tell me, when the moon is nearly full, so that it can be seen by day and also in the middle of the night, does it appear more brilliant in the daytime or at night?

SIMPLICIO: Incomparably more at night. . . . Thus I have observed the moon by day sometimes among small clouds, and it looked like a little bleached one; but on the following night it shone most splendidly.

SALVIATI: So that if you had never happened to see the moon except by day, you would not have judged it brighter than one of those little clouds. . . . Now tell me, do you believe that the moon is really brighter at night than by day, or just that by some accident it looks that way?

SIMPLICIO: I believe that it shines intrinsically as much by day as by night. . . .

SALVIATI: Now, have you ever seen the earth lit by the sun in the middle of the night? . . .

SIMPLICIO: It is impossible for anyone who is on the earth, as we are, to see by night that part of the earth where it is day. . . .

SALVIATI: So you have never chanced to see the earth illuminated except by day, but you see the moon shining in the sky on the darkest night as well. And that, Simplicio, is the reason for your believing that the earth does not shine like the moon, for if you could see the earth illuminated while you were in a place as dark as night, it would look to you more splendid than the moon. So, if you want to proceed properly with the comparison, we must draw our parallel between the earth's light and that of the moon as seen in daytime, not that of the nocturnal moon. . . .

Now you yourself have already admitted having seen the moon by day among little whitish clouds, and similar in appearance to one of them. This amounts to granting at the outset that these little clouds, though made of elemental matter, are just as fit to receive light as the moon is. More so, if you will recall in memory having seen some very large clouds at times, white as snow. It cannot be doubted that if such a cloud could remain as luminous on the darkest night, it would light up the surrounding region more than a hundred moons.

If we were sure, then, that the earth is as much lighted by the sun as one of those clouds, no question would remain about its being no less brilliant than the moon. Now, all doubt on this point ceases when we see those same clouds, in the absence of the sun, remaining as dark as the earth all night long. And what is more, there is not one of us who has not seen such a cloud low and far off, and wondered whether it was a cloud or a mountain – a clear indication that mountains are no less luminous than those clouds. (D 87–9)

The need to keep his readers' minds on their own visual

experiences in order that his kind of science might prevail over the verbalism of natural philosophy was clear to Galileo, and he managed this by devices he had learned from the poets. He was a great admirer of poetry and ascribed his own literary style to study of Ariosto, whose works he is said to have known by heart. In the comet controversy of 1619–23, his Jesuit adversary had appealed to the testimony of poets to prove the existence of certain phenomena, which Galileo had ridiculed as evidence, saying that Nature takes no delight in poetry. It was quite a different matter to use poetic tricks of description, which served not to invoke the authority of some author but to make vivid the reader's awareness of things seen. During that same period Galileo had given encouragement to some young poets who were endeavouring to do for literature the kind of thing Galileo's father had done for music and he himself was doing for science. He had also written a detailed comparison of the poetry of Ariosto and Tasso. In carrying his appeal for the new science to laymen, Galileo used literary devices as a tool no less than he used mathematics to convey it to colleagues.

The same interest in language appeared in Galileo's argument against theological interference in freedom of scientific inquiry. In his *Letter to Christina* he stressed the position that the writers of the Bible, with divine inspiration, employed language that would be easily understood by ordinary people and would not raise doubts in their minds about matters of faith and salvation by speaking of the stability of the sun and motion of the earth. They foresaw, in Galileo's view, that if they taught such things literally, people who doubted them would also question their correctness on matters of more serious import which it was their main purpose to impart. Bellarmine understood this, but warned against the use of such arguments by Galileo, because it would infuriate less wise opponents.

Not only semantic considerations, but distinctions between questions answerable by science and those belonging to philosophers only were made by Galileo. This he wrote to Liceti:

> The problem or question of the centre of the universe, and whether the earth is situated there, is among the least worthy of consideration in the whole of astronomy. It has sufficed the greatest astronomers to assume that the terrestrial globe is of insensible size

in comparison with the starry orb, and that as to location it is either at the centre of the diurnal revolution of that orb or is removed therefrom by an insignificant distance. There is no reason to tire oneself out trying to prove that, nor that the fixed stars are situated in a space bounded by a spherical surface; it is enough that they are located at an immense distance from us. Likewise, to want to assign a centre to that space, of which the shape is neither known nor can be known (or even whether it has a shape), is in my opinion a superfluous and an idle task. To believe that the earth can be located at a centre not known to exist in the universe is indeed a frustrating enterprise. (GW 411)

Such, to Galileo, were the traditional enterprises of natural philosophers, from which one should turn to useful science. Until his time science had been the handmaiden of philosophy, which in turn was the handmaiden of theology. He wished to free science from subservience to philosophy as the historical obstacle to its utility and progress. He dreamed of a better philosophy as the ultimate result, but the absolutist conception that science should be free from all constraint was no part of Galileo's thought. Its constraints should be chosen in such a way that no conflict with theology could ever arise, and out of it would eventually come a philosophy equally in harmony with it and with theology.

A more radical conception of science as demanding absolute freedom was later formulated by intellectuals who saw religion as an uncompromising foe of science. Probably that would not have gained much hold if the Catholic Church had not made the disastrous error – which Galileo had done his best to prevent – of taking an official position for one kind of science against another. Once it took that position, there ensued the usual long period of attempting to justify a mistake by rigid adherence to it. Eventually, in 1890, the Church shifted to a position much like that suggested by Galileo in his *Letter to Christina.* While the present book was being edited for publication, Pope John Paul II recognised Galileo's *Letter* as having 'formulated important norms of an epistemological character, which are indispensable to reconcile Holy Scripture and science.' There is no reason to think that that was not his intention when he wrote the *Letter.*

Galileo felt crushed by the verdict of 'vehement suspicion of heresy' because it cut him off from the church he loved; because he knew that no heresy had ever crossed his mind; because the verdict was a second error by an institution to which men looked for truth, and because his own life work stood condemned. He was not disheartened because he considered the Copernican astronomy, or anything else in science as he saw it, to be absolutely and completely true. That is evident from a letter he dictated to a friend in 1641, the last year of his life, with no hope of reward or any fear of further punishment:

> The falsity of the Copernican system must not on any account be doubted, especially by us Catholics, who have the irrefragable authority of Holy Scripture interpreted by the greatest masters in theology, whose agreement renders us certain of the stability of the earth and the mobility of the sun around it. The conjectures of Copernicus and his followers offered to the contrary are all removed by that most sound argument taken from the omnipotence of God. He being able to do in many, or rather in infinite ways, what to our view and observation seems to be done in one particular way, we must not pretend to hinder God's hand and stubbornly maintain that in which we may be mistaken.
>
> And just as I deem the Copernican observations and conjectures inadequate, so I judge equally, and more, fallacious and erroneous those of Ptolemy, Aristotle, and their followers, when without going beyond the bounds of human reasoning, their inconclusiveness can be very easily discovered. (GW 417)

To reject all systems is to say, as Galileo had said in his *Dialogue*: 'There is no event in Nature, not even the least that exists, such that it will ever be completely understood by theorists.' Galileo was as deeply loyal to science 'as a method of reasoning capable of human pursuit' as any man who ever lived, but the kind of science he proposed was quite different from that which supposes any problem to have been completely and finally solved. Science as a method of successive approximations cannot be beaten as an unending search for truth about the universe. How it is possible to search for something one knows can never be reached, and yet be sure that each year one is closer than last year, is a philosophical and not a scientific question. The scientist's answer is in the theory and practice of measurement and in the theory of error, by which precision of measurement may be judged without one's knowing the precise measurement absolutely.

Galileo's own conscience was clear both as Catholic and as scientist. On one occasion he wrote, almost in despair, that at times he felt like burning all his work in science; but he never so much as thought of turning his back on his faith. The Church turned its back on Galileo, and has suffered not a little for having done so; Galileo blamed only some wrong-headed individuals in the Church for that. When Nicole Fabri de Peiresc, an eminent French amateur of science who had heard Galileo lecture at Padua in 1603, wrote to tell him that he was writing to the authorities at Rome demanding Galileo's pardon, Galileo replied:

> Your Excellency's letter, filled with feelings of courtesy and good will, continues to make the fortune of my misfortune appear sweeter to me, and in a certain way to bless the persecutions of my enemies, without which there would have remained hidden from me that which is most to be admired in humanity, and the benign inclination of my noble patrons, and above all your Excellency's love. . . . They are moved to compassion for my situation, in which, in addition to the reason named, there is for me no little comfort in believing that it is not a spirit of ever-increasing cruelty that continues to hold me under oppression, but rather, as I shall say, a sort of official policy on the part of those who want to cover up their original error of having wronged an innocent man by continuing their offences and wrongs, so that people will conceive that other grave demerits, not made public, may exist to aggravate the guilt of the culprit. (L 52)

And again, later, to the same correspondent:

> I have said that I hope for no relief, and this is because I committed no crime. If I had erred, I might hope to obtain grace and pardon; for transgressions by subjects are the means by which the prince finds occasion for the exercise of mercy and indulgence. Hence when a man is wrongly condemned to punishment, it becomes necessary for his judges to use greater severity in order to cover up their own misapplication of the law. This afflicts me less than people may think possible, for I have two sources of perpetual comfort – first, that in my writings there cannot be found the faintest shadow of irreverence towards the Holy Church; and second, the testimony of my own conscience, which only I and God in Heaven thoroughly

know. And He knows that in this cause for which I suffer, though many might have spoken with more learning, none, not even the ancient Fathers, have spoken with more piety or with greater zeal for the Church than I. (PLG 278–9)

The cause for which Galileo suffered, in his own view, was clearly not Copernicanism but sound theology and Christian zeal. That 'misapplication of law' to which Galileo referred can hardly have been his condemnation in 1633, which so far as he was concerned was an error of fact. What grieved Galileo was the theologians' error of 1616, as an indirect result of which he had been punished. Their error was in his eyes a misapplication of law established by the ancient Fathers who had wisely separated science from religion.

Galileo went on to say that if the frauds and stratagems that had been used at Rome in 1616 to impose upon the supreme authority could be revealed, the uprightness of his intentions would be clear. Since theologians *were* the supreme authority, the frauds and stratagems by which they had been imposed on must have come from other men, identified in the *Letter to Christina* as professors of philosophy. Similarly, by 'uprightness of intentions' Galileo cannot refer to support of Copernicus, but only to his campaign for freedom of scientific inquiry without Church intervention.

Galileo died with a clear conscience at Arcetri on 9 January 1642. A few days later Luke Holste, who was attached to the household of Francesco Cardinal Barberini, the most important of the three who withheld their signatures from Galileo's condemnation, wrote to a friend at Florence:

Today news has come of the loss of Signor Galilei, which touches not just Florence but the whole world, and our whole century which from this divine man has received more splendour than from almost all the other ordinary philosophers. Now, envy ceasing, the sublimity of that intellect will begin to be known which will serve all posterity as guide in the search for truth. (GW 436)

Reading list

The literature on Galileo is vast and in many languages. His works and correspondence are available in *Le Opere di Galileo Galilei*, edited by Antonio Favaro and published at Florence, 1890–1910, reprinted with additions in later years. Bibliographies were compiled by A. Carli and Favaro (1588–1895) and by G. Boffito (1896–1940), continued to 1964 by E. McMullin in *Galileo Man of Science* (see below). The following list is limited to books in English; important journal articles and monographs after 1964 will be found by consulting bibliographies and notes in recent books, while earlier articles are identified in the three major bibliographies. English books of minor importance, and those dealing with Galileo only as a part of some other theme, have been omitted.

A. Works left unpublished by Galileo

W. A. Wallace: *Galileo's Early Notebooks: The Physical Questions* (Univ. of Notre Dame, 1977)

I. E. Drabkin and S. Drake: *Galileo On Motion and On Mechanics* (Univ. of Wisconsin, 1960)

S. Drake and I. E. Drabkin: *Mechanics in Sixteenth-Century Italy* (Univ. of Wisconsin, 1969)

B. Published books by or attributed to Galileo, in order of date

1605 Cecco di Ronchitti, *Dialogo ... della stella nuova* (Padua): trans. in S. Drake, *Galileo Against the Philosophers* (Zeitlin & Ver Brugge, Los Angeles, 1976)

1606 *Le Operazioni del compasso ...* (Padua): trans. S. Drake, *Operations of the Geometric and Military Compass* (Smithsonian Institution, Washington, 1978)

1610 *Sidereus Nuncius ...* (Venice): *The Starry Messenger*, trans. and abridged in S. Drake, *Discoveries and Opinions of Galileo* (Doubleday, New York, 1957), cited hereinafter as *Discoveries*

1612 *Discorso intorno alle cose ... in su l'acqua ...* (Florence): trans. Thomas Salusbury, ed. S. Drake, *Discourse on Bodies in Water* (Univ. of Illinois, 1960)
1613 *Istoria ... delle macchie solari* (Rome): trans. *Letters on Sunspots*, abridged in *Discoveries*
1615 *Lettera a Madama Cristina de Lorena* (Strasbourg, 1636): trans. *Letter to the Grand Duchess Christina*, in *Discoveries*
1619 Mario Guiducci, *Discorso delle comete* (Florence): trans. *Discourse on the Comets* in S. Drake and C. D. O'Malley, *Controversy on the Comets of 1618* (Univ. of Pennsylvania, 1960), hereinafter cited as *Controversy*
1623 *Il Saggiatore ...* (Rome): trans. in *Controversy*; also abridged in *Discoveries*
1632 *Dialogo ...* (Florence): trans. S. Drake, *Dialogue Concerning the Two Chief World Systems* (Univ. of California, 1953, rev. 1967)
1638 *Discorsi ... intorno a due nuove scienze ...* (Leyden): trans. S. Drake, *Two New Sciences* (Madison, 1974)

C. Biographies

1829–30 [J. E. Drinkwater], *Life of Galileo* (London)
1870 [Mary Allan-Olney], *The Private Life of Galileo* (London)
1879 Karl von Gebler (trans. Sturge), *Galileo Galilei and the Roman Curia* (London)
1903 J. J. Fahie, *Galileo, His Life and Work* (London)
1931 Emil Namer (trans. Harris), *Galileo* (New York)
1938 F. S. Taylor, *Galileo and the Freedom of Thought* (London)
1955 G. de Santillana, *The Crime of Galileo* (New York)
1964 James Brodrick, S.J., *Galileo: The Man, his Work, his Misfortunes* (London)
1965 Ludovico Geymonat (trans. Drake), *Galileo Galilei: A biography and inquiry into his philosophy of science* (New York)
1978 Stillman Drake, *Galileo At Work: His Scientific Biography* (Chicago)

D. Special Studies or Collected Essays

1965 M. Kaplon, ed., *Homage to Galileo* (Cambridge, Mass.)

1966 C. Golino, ed., *Galileo Reappraised* (Berkeley and Los Angeles)

1967 E. McMullin, ed., *Galileo Man of Science* (New York)

1970 Stillman Drake, *Galileo Studies* (Ann Arbor, Mich.)

1971 J. J. Langford, *Galileo, Science, and the Church* (Ann Arbor, Mich.)

1972 W. R. Shea, *Galileo's Intellectual Revolution* (New York)

1974 Maurice Clavelin (trans. Pomerans), *The Natural Philosophy of Galileo* (Cambridge, Mass.)

1974 Dudley Shapere, *Galileo: A Philosophical Study* (Chicago)

1978 Alexandre Koyré (trans. Mepham), *Galileo Studies* (Hassocks, Sussex)

1978 R. E. Butts and J. C. Pitt, eds., *New Perspectives on Galileo* (Boston)

Index

Past Masters

AQUINAS Anthony Kenny

Anthony Kenny writes about Thomas Aquinas as a philosopher, for readers who may not share Aquinas's theological interests and beliefs. He begins with an account of Aquinas's life and works, and assesses his importance for contemporary philosophy. The book is completed by more detailed examinations of Aquinas's metaphysical system and his philosophy of mind.

FRANCIS BACON Anthony Quinton

Francis Bacon is a major figure in the history of thought, and England's most important Renaissance thinker. His major works – *The Advancement of Learning* and *Novum organum* – are philosophical, and it is these which Anthony Quinton takes as his central topic. He also gives a brief account of Bacon's life, and discusses his other writings (including the *Essays*), his intellectual environment, and the influence of his work.

BURKE C. B. Macpherson

This new appreciation of Edmund Burke introduces the whole range of his thought, and offers a novel solution to the main problems it poses. Interpretations of Burke's ideas, which were never systematised in a single work, have varied between apparently incompatible extremes. C. B. Macpherson finds the key to an underlying consistency in Burke's political economy which, he argues, is a constant factor in Burke's political reasoning.

DANTE George Holmes

George Holmes expresses Dante's powerful originality by identifying the unexpected connections the poet made between received ideas and his own experience. He presents Dante's biography both as an expression of the intellectual dilemma of early Renaissance Florence and as an explanation of the poetic, philosophical and religious themes developed in his works. He ends with a discussion of the *Divine Comedy*, Dante's poetic panorama of hell, purgatory and heaven.

HOMER Jasper Griffin

The ideas of the *Iliad* and the *Odyssey* have had an incalculable influence upon the thought and literature of the West, yet the two epics are very different. It is the greatness and the fragility of Achilles which define the human condition, whereas Odysseus represents a new heroism of endurance and guile. Jasper Griffin shows how each epic has its own coherent and suggestive view of the world, and of man's place within it.

HUME A. J. Ayer

A. J. Ayer begins his study of Hume's philosophy with a general account of Hume's life and works, and then discusses his philosophical aims and methods, his theories of perception and self-identity, his analysis of causation, and his treatment of morals, politics and religion. He argues that Hume's discovery of the basis of causality and his demolition of natural theology were his greatest philosophical achievements.